Combat Readiness Through Medicine at the Battle of Antietam

THE HUMAN FACE OF OUR BLOODIEST DAY

SCOTT C. WOODARD, MA
GEORGE C. WUNDERLICH, BA
WAYNE R. AUSTERMAN, PhD

BORDEN INSTITUTE
US ARMY MEDICAL CENTER OF EXCELLENCE
FORT SAM HOUSTON, TEXAS

Edward A. Lindeke | Colonel (Retired), MS, US Army | Director, Borden Institute
Gina D. Frank | Senior Production Editor, Borden Institute
Richard P. Salomon | Volume Editor, Borden Institute
Christine Gamboa-Onrubia, MBA | Book Design, Fineline Graphics, LLC
Robert A. Dredden | Illustrator
Ernest J. Barner | Public Affairs

The opinions or assertions contained herein are the personal views of the author and are not to be construed as doctrine of the Department of the Army or the Department of Defense. Use of trade or brand names in this publication does not imply endorsement by the Department of Defense.

CERTAIN PARTS OF THIS PUBLICATION PERTAIN TO COPYRIGHT RESTRICTIONS. ALL RIGHTS RESERVED. NO COPYRIGHTED PARTS OF THIS PUBLICATION MAY BE REPRODUCED OR TRANSMITTED IN ANY FORM OR BY ANY MEANS, ELECTRONIC OR MECHANICAL (INCLUDING PHOTOCOPY, RECORDING, OR ANY INFORMATION STORAGE AND RETRIEVAL SYSTEM), WITHOUT PERMISSION IN WRITING FROM THE PUBLISHER OR COPYRIGHT OWNER.

Published by Borden Institute
US Army Medical Center of Excellence, Fort Sam Houston, Texas

Library of Congress Cataloging-in-Publication Data

Names: Woodard, Scott C., author. | Wunderlich, George C, author. | Austerman, Wayne R. (Wayne Randolph), 1948- author.
Title: Combat readiness through medicine at the Battle of Antietam : the human face of our bloodiest day / Scott C Woodard, George C Wunderlich, Wayne R Austerman.
Description: Fort Sam Houston, Texas : Borden Institute, [2022] | Includes bibliographical references and index. | Summary: "This publication highlights important medical innovations and improvements gained from the deadliest day in US history, the Battle of Antietam. This pivotal US Civil War battle helped shape future combat medical readiness practices in the US Army. The Battle of Antietam provides important lessons in battlefield tactics, leadership, command and control, communications, and unit training that improve the nation's readiness to bring combat power to commanders in the field of battle. It was during this battle that the US Army solidified its emerging plan to decisively combat battlefield mortality, which marked the beginning of true combat readiness through medicine. This publication is applicable to the entire range of health care in the Department of Defense and can serve as a valuable learning aid for a variety of military and civilian medical professionals. This study can be used in conjunction with the US Army Center of Military History's Staff Ride Guide: Battle of Antietam or it can be used separately as a focused analysis of military medicine"—Provided by publisher.
Identifiers: LCCN 2022034450 (print) | LCCN 2022034451 (ebook) | ISBN 9781737131120 (hardcover)
Subjects: LCSH: Antietam, Battle of, Md., 1862. | United States—History--Civil War, 1861-1865—Medical care.
Classification: LCC E474.35 .W66 2022 (print) | LCC E474.65 (ebook) | DDC 973.7/336—dc23/eng/20220727
LC record available at https://lccn.loc.gov/2022034450
LC ebook record available at https://lccn.loc.gov/2022034451

Printed in the United States of America
26 25 24 23 22   10 9 8 7 6 5 4 3 2 1

CONTENTS

Foreword | vii
About the Authors | viii
Dedications | x
Acknowledgments | xi
Preface | xii
Introduction | xv

Read Ahead | 1
CHAPTER 1 Direct Action on the Battlefield | 21
    Stop 1: North Woods | 21
    Stop 2: D.R. Miller's Cornfield | 27
    Stop 3: Dunker Church | 36

Read Ahead | 45
CHAPTER 2 Direct Action and Battlefield Hospital | 53
    Stop 1: Clara Barton | 53
    Stop 2: Sunken Road (Bloody Lane) | 61
    Stop 3: Straw-Stack Hospital | 67
    Stop 4: Roulette Barn | 75

Read Ahead | 84
CHAPTER 3 Last Push | 95
    Stop 1: Burnside Bridge | 95
    Stop 2: Otto and Sherrick Farm Hospitals | 110
    Stop 3: High Point | 112

CHAPTER 4 Division-Level Field Hospitals | 117
    Stop 1: Philip Pry House Field Hospital Museum | 117
    Stop 2: Samuel Pry Mill Hospital (Valley Mills) | 127
    Stop 3: Dr Otho J. Smith's Farm (French's Division) Hospital | 129
    Stop 4: Keedysville Hospitals | 131
    Stop 5: German Reformed Church Hospital
             (Mount Vernon Reformed Church) | 139

CHAPTER 5 Corps and Evacuation Hospitals | 149
    Stop 1: Main Supply Route | 149
    Stop 2: Zion Lutheran Church Hospital | 153
    Stop 3: German Reformed Church Hospital
           (Christ Reformed United Church of Christ) | 154

CHAPTER 6 Frederick General Hospitals and the
              National Museum of Civil War Medicine | 157
    Stop 1: General Hospital Number 5, Visitation Academy | 161
    Stop 2: General Hospital Number 4 | 168
    Stop 3: General Hospital Number 3 | 176
    Stop 4: National Museum of Civil War Medicine | 187

Selected Antietam Medical Histories | 193
Antietam Casualties | 200
Medical Officers of the Army of the Potomac at Antietam | 202
Index | 203

Combat Readiness Through Medicine at the Battle of Antietam

THE HUMAN FACE OF OUR
BLOODIEST DAY

Map of the Battle of Antietam. Robert Knox Sneden was a soldier and mapmaker in the Union Army of the Potomac. His work captured astonishing details wherever he traveled. This watercolor and pen-and-ink map places the locations of the battling Union and Confederate armies opposing one another near Sharpsburg, Maryland, on 17 September 1862. Source: Library of Congress.

FOREWORD

Learning from our collective past is a useful return on investment, if that investment is thoughtfully applied to help make us better. Standard US Army professional development incorporates staff ride procedures in its lexicon, but this work expands on that genre by pulling out the medical aspects of a battle. Additionally, we know this publication is applicable to the entire range of health care in the Department of Defense and beyond. Members of all military service medical departments, whether a ward nurse at a fixed facility or a young US Army brigade surgeon in a ground combat unit, can benefit from the experience our forefathers earned with sweat and blood. Whether the reader is a Navy Corpsman, Air Force Flight Surgeon, community fire chief, or a regional trauma center emergency room technician, the story of the emerging modern medicinal system used on the battlefield is a valuable learning aid for them and for others outside the military.

This publication serves to highlight important medical innovations and improvements gained from the deadliest day in US history. The Battle of Antietam in the US Civil War was pivotal in shaping future combat medical readiness practices in the US Army. The medical aspects of the Battle of Antietam provide important lessons in battlefield tactics, leadership, command and control, communications, and unit training that improve our readiness to bring combat power to commanders in the field of battle.

Raymond S. Dingle
Lieutenant General, US Army
The Surgeon General and
Commanding General, USAMEDCOM

ABOUT THE AUTHORS

WAYNE R. AUSTERMAN, PhD, is a retired historian who served at the Department of Leader Training at the US Army Medical Center of Excellence for 21 years. A veteran of 5 years of active duty as a US Army infantry officer in the old 2nd Armored Division, Austerman received his BA (1971), MA (1977), and PhD (1981) from Louisiana State University in Baton Rouge. He has directed the staff ride program at the Department of Leader Training since his first foray to the Antietam National Battlefield at Sharpsburg, Maryland, in 1990. Austerman published four articles in the *US Army Medical Department Journal* and the book *Musket and Arrow: a Guide to US Army Staff Ride Sites in Southern Texas* (US Army Medical Department Center and School). His other major works include *Sharps Rifles and Spanish Mules: the San Antonio–El Paso Mail, 1851–1881* (Texas A&M University Press); *Program 437: The USAF's First Anti-Satellite Weapons System* (Colorado Springs: Space Command History Office); as well as over a hundred other articles in a diverse array of journals.

SCOTT C. WOODARD, MA, is a historian at the US Army Medical Department Center of History and Heritage, Medical Center of Excellence, and previously served in the US Army for over 22 years. He holds a bachelor's in history from The Citadel, The Military College of South Carolina, and a master's in military medical history from the Uniformed Services University of the Health Sciences at Bethesda, Maryland. "Woody" is a certified Military Historian from the US Army Center of Military History. He has published articles in numerous periodicals including *Military Review, Task and Purpose, Military Medicine, Army Logistician,* and *Army Times*. Most recently, Woody's chapter, "Forward Surgery in the Korean War: The Mobile Army Surgical Hospitals," was published in *The Evolution of Forward Surgery in the US Army: From the Revolutionary War to the Combat Operations of the 21st Century* by the Borden Institute in 2018.

GEORGE C. WUNDERLICH, BA, is the Director of the US Army Medical Department Museum. Wunderlich previously served for 13 years as Executive Director of the National Museum of Civil War Medicine and its two subsidiary museums: the Pry House Field Hospital Museum and the Clara Barton Missing Soldiers Office Museum. From 2004 to 2015, Wunderlich provided staff rides at the Antietam National Battlefield for personnel at the Joint Medical Executive Skills Institute Capstone Symposium, US Army Medical Material Agency, and the Uniformed Services University of the Health Sciences. He also served as Historical Coordinator of the Presidential Management Fellows Program Antietam Staff Ride at the Eastern Management Development Center for the Office of Personnel Management from 2002 to 2006. Since 1999, Wunderlich has provided historical consulting services to more than 75 television productions in Great Britain, Canada, and the United States.

DEDICATIONS

This volume is dedicated to my great-grandfather, Confederate Private William Patrick Goode, Company B, 57th Virginia Infantry, Armistead's Brigade, Pickett's Division, Longstreet's Corps, Army of Northern Virginia. While attached to Confederate General A.P. Hill's Division at the fall of Harper's Ferry, West Virginia, the 57th Virginia marched swiftly northward to rejoin General Robert E. Lee's army, just in time to smash the Union IX Corps' attack on Lee's right on the afternoon of 17 September 1862. —*Wayne R. Austerman*

This work is dedicated to my 3rd great-uncles, Confederate Privates Thomas Woodard and Josiah Woodard of the 24th North Carolina Infantry, Ransom's Brigade, Walker's Division, Longstreet's Corps, Army of Northern Virginia. After taking Harper's Ferry, the infantry fought in the West Woods near the Dunker Church and on the high ground surrounding the Alfred Poffenberger Farm during the Battle of Antietam at Sharpsburg, Maryland. —*Scott C. Woodard*

This work is dedicated to my 2nd great-grandfather, Union Private William Wallace Pedrick of the 13th Massachusetts Infantry, First Corps, Second Division, who marched into D.R. Miller's Cornfield at the Battle of Antietam and saw almost half of his regiment killed or wounded in 30 minutes. It is also dedicated to my wife Irene B. Wunderlich, EdD. Without her untiring support and counsel, my accomplishments over the past 30 years would have been impossible. —*George C. Wunderlich*

ACKNOWLEDGMENTS

I am eternally grateful for the inspiration and encouragement from countless friends. A select group of people should be specifically noted in thanks: Jake Wynn of the National Museum of Civil War Medicine in Frederick, Maryland, for introducing me to Charles Johnson and *The Long Roll*; Terry Reimer, also from the National Museum of Civil War Medicine, for sharing maps and manuscripts; David Price, another great representative of the National Museum of Civil War Medicine, for his advice on Frederick history and medicinal libations; Elizabeth Howe from the John Clinton Frye Western Maryland Room in Hagerstown, Maryland, for finding the elusive George Allen story in *The Antietam Wavelet* newspaper; Laura Elliot's post about Confederate Private George Washington Lafayette Ard on *CivilWarTalk* (https://civilwartalk.com/); John Bank's many ideas and inspirations on *John Banks' Civil War Blog* (http://john-banks.blogspot.com); David Wallace and the wonderful staff and volunteers at the All Saints' Episcopal Church in Frederick for their gracious tour and resources; George Wunderlich, Director of the US Army Medical Department Museum, and brother, for his enthusiasm and rich knowledge of Civil War medicine; and Wayne Austerman for his over 20-year friendship and inspiration to create whimsical titles. And finally, I would like to thank my editor Richard P. Salomon. His questions (he was a US Navy public affairs specialist), great eye, and suggestions made for a much better product. Mistakes in the historical narratives and citations are my error.

Scott C. Woodard
Historian
US Army Medical Department
Center of History and Heritage

PREFACE

This book focuses on the medical aspects of the single bloodiest day in combat for the United States of America. It was here that the Medical Director of the Army of the Potomac, Major (Dr) Jonathan Letterman, solidified his emerging plan to decisively combat battlefield mortality and mark the beginning of true combat readiness through medicine. It is from this crucible that the US Army Medical Department began its path toward a world-renowned reputation for rendering aid to those most dear to the combat mission—the soldier. It is from this battlefield that we see the human dimension and usefulness of Dr Letterman's plan for future generations in joint interoperability, standardization, and evidence-based medicine.

Participants can further examine the story of battlefield medicine in the rich outdoor classroom of western Maryland by using the 2007 US Army Center of Military History (CMH) *Staff Ride Guide: Battle of Antietam* by Ted Ballard. This publication serves as the core text to help participants understand the contextual, operational, and tactical overview of the battle. Traveling back in time, participants begin the staff ride at D.R. Miller's Cornfield and Dunker Church in Sharpsburg, Maryland, where major action took place. After that, they will encounter division-level battlefield hospitals, travel through the corps-level evacuation and treatment route, and then complete the journey in Frederick, Maryland, at the general hospitals and at the National Museum of Civil War Medicine.

This study may be used in conjunction with the previously published CMH guide, or it may be used separately as a focused analysis of military medicine. Additionally, the points of interest, or stops, may be grouped into the physical battlefield (Chapters 1 through 3), division-level hospitals (Chapter 4), and the large rear-area hospitals and National Museum of Civil War Medicine (Chapters 5 and 6). Each of these groupings may be studied together or separately from one another.

So, what is an Army staff ride? In 1906, Major Eben Swift, of the US Infantry and Cavalry School at Fort Leavenworth, Kansas, led officers from Fort Leavenworth to a nearby American Civil War battlefield and conducted the first "riding" of staff members through a national military park.[1] This mode of learning was adopted into the professional curriculum for Army officers and has since been adopted by other US military branches, the National Park Service, and civilian historians worldwide. The original national military parks were battlefields from the US Civil War that were preserved to instruct Army officers in the art of war. The insights gained from walking the terrain and reaching into the past to experience the fog of war with modern lenses is an enriching vicarious experience not duplicated anywhere else. Today, at the Antietam National Battlefield, students of history can still see the plaques that were placed along the battlefield by the War Department around 1897, which document the movements the combatants made. These markers provided early visual aids for Army officers who wanted to apply lessons from the battle.

It is fitting that Army Medicine is adopting this technique of envisioning the battle from the perspective of medicine. Major General Eben Swift's father was Army Surgeon Brevet Brigadier General Ebenezer Swift. From 1847 to 1883, Brevet Brigadier General Swift served in the Mexican–American War, Civil War, Indian Wars, and on the Texas frontier in peacetime. The role of the US Army soldier and US Army Medicine continues to intertwine. This "Medical Annex" to the *Staff Ride Guide: Battle of Antietam* serves to continue that legacy.

                              Scott C. Woodard
                              Historian
                              US Army Medical Department
                              Center of History and Heritage

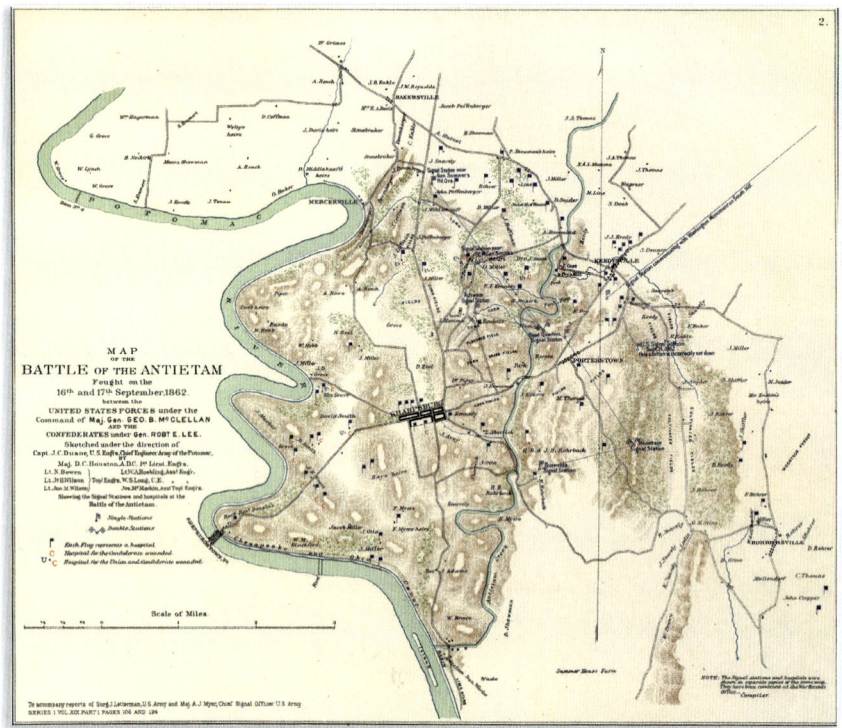

Antietam Map. Solid flags represent hospital locations. The "U" represents wounded Union soldiers and the "C" represents wounded Confederate soldiers. Both sides treated enemy patients at their respective hospitals; treatment was not confined to "friendly" soldiers. Source: Davis GB, Cowles CD, and Caldwell JA. US War Department. New York, NY: Atlas Publishing Co; 1892.

* Unless otherwise noted, the letters, tables, and other graphics in this book are for illustration purposes only and are not reproductions of the original documents.

INTRODUCTION

On a freezing December day in 2003, Navy Commander Lori Frank, retired Army Colonel Richard Shipley, and I stood at the edge of D.R. Miller's Cornfield looking south at the Smoketown Road and tried to envision the approach of the Texas Brigade as it came north to push the Federal forces away from Confederate General Robert E. Lee's left flank. Despite the bitter wind, the clear day gave a picture perfect view of the landscape. We were planning a March 2004 staff ride for members of the Joint Medical Executive Skills Institute and, on this day, the grand assault of the Texans, and indeed the entire battle, was only a backdrop for an even larger story—that of medical innovation on the battlefield. This staff ride had already been 4 years in the making, and looking northeast toward the Washington Monument on South Mountain, we could all clearly see the very spot that inspired this effort.

In the early spring of 1999, a freak snow squall hit the area along the Maryland-Pennsylvania border. At 3 am, the alarms sounded in four firehouses for a serious personal injury accident near the top of South Mountain, about 7 miles northeast of the Antietam National Battlefield at Sharpsburg, Maryland. The snow was heavy and wet, and visibility had fallen to near zero. It was my night to drive the Myersville Volunteer Fire Company Ambulance 89. Our chief arrived at the scene first and found that there were six vehicles involved with at least seven people seriously injured, including one victim who had been ejected from her vehicle. Within a few minutes, the Boonsboro, Mount Aetna, and Hagerstown units arrived on scene, and the evacuation began in earnest.

As I watched the scene unfold from my perch on the running board of my ambulance, I was struck by the scope of the scene. I was tasked that morning to keep a safety watch over my emergency medical technicians and bring them needed supplies before delivering the patients to Washington County Hospital's Trauma Care Center. There would be no aviation this day. Poor visibility prevented that. Our priority-one patient was the ejected victim. She had not only been ejected through a rear window of a van, but she had spent too many minutes on the cold ground, in the snow, and was in shock. Her ride to the hospital would require covering the next 8 miles on unplowed interstate in near white-out conditions. Time was not on her side and road conditions were not on ours. As I drove, something kept nagging at me. There was something about this call that was familiar, but I could not place why. I did not have time to think about it. Driving required my full attention.

When we reached the trauma center, she was still with us. After handing our patient off to the waiting trauma team, my medics restocked our supplies, and I cleaned the patient bay. After that, we headed home. As we left the hospital, the ground still had 5 to 6 inches of new snow that had fallen in less than 2 hours—but it was melting fast. The storm passed as quickly as it came. On the way home, we marveled at the beauty of the now moonlit landscape. I reflected back and wondered what had seemed so familiar about this call. I then looked up and saw the Washington Monument on the top of South Mountain, and it hit me like a brick. The call that our team had just answered took place about 2 miles from the battlefield where the ambulance system developed by Major (Dr) Jonathan Letterman had its first test of fire under his direction. The hair stood up on the back of my neck. I realized that nearly every action we took during that accident call had been touched, in some way, by the battles of South Mountain and Antietam nearly 150 years before.

Four years after that call, standing at Miller's Cornfield where my own great-great grandfather fought, I gazed up at the Washington Monument again and thought back to that night. After that ambulance call, I spent an entire year studying Dr Letterman and his medical orders of August and October 1862. That research turned into a series of lectures, a staff ride, and an institute under the direction of the National Museum of Civil War Medicine. In 2015, retired Army Major Scott Woodard, Dr Marble Sanders, and Army Major Ken Koyle, all under the US Army Medical Department Center of History and Heritage, joined the effort to develop a medical staff ride that would be in line with Army learning standards. You are holding the result of all these efforts in your hand.

George C. Wunderlich
Director
US Army Medical Department Museum

Civil War Battlefield Medicine: An Acute Overview

The professional Army of the United States, first formed in 1775, was selectively small and reflected the founding generation's fear of a large, standing Army. Even though war had called upon the medical services to repair the wounded and assuage the fevered brows of soldiers in the War of Independence, the War of 1812, and the Mexican–American War, the cadre of medical personnel serving in the US Army was minimal. The thousands of wounded men and animals left dying on the battlefield near Manassas, Virginia, during the First Battle of Bull Run in 1861 gave testimony to the unpreparedness facing both armies. Both armies and their medical departments were not ready. Using lessons learned about sanitation in the Crimean War, civilian relief agencies, such as the US Christian Commission and the US Sanitary Commission, began to fill in the deadly gaps exposed on the bloody fields of combat.

The mid-19th century marked a shift in medicine. The long-respected Greek medical teachings of "balancing the humors" (a theory that stated good health came from the proper balance of bodily fluids) to maintain health, and the slowly emerging scientific method that would eventually prove that germs were not just a theory and anesthesia benefitted the patient, all began to converge during this tumultuous time. Medical officers of the Regular Army were subjected to vigorous boards of review and reflected the best of medical knowledge of their time. But as the flash of battle struck down the men in wool, the complications of logistics and administration structured for peacetime became unbearable and broke under the weight of exponential growth and rapid mobilization of citizens to soldiers.

After his appointment as the Medical Director of the Army of the Potomac under Union General George B. McClellan in July 1862, Major (Dr) Jonathan Letterman began working with McClellan and Brigadier General (Dr) William Hammond, the Surgeon General of the Army, to implement a new method of battlefield evacuation and treatment. This was the turning point in medical readiness and, therefore, combat readiness in the US Army. Dr Letterman devised an ambulance corps of soldiers solely directed by the medical department and tied into a predesignated hospitalization plan. Before this command-directed order, the corps and regimental ambulances were controlled by the Quartermaster Department and were often operated by ill-disciplined civilian contractors. Through Dr Letterman's specified tasks of weekly inspections and drills, the days of broken ambulances and untrained soldiers were over. Before Dr Letterman's reforms, insular decisions were made for field hospital locations. Some surgeons refused to treat members of other regiments and had little idea of the location of other hospitals. This created a disconnected system where patient flow and surgical intervention should meet. The health of the Army and, therefore, its fighting strength were in jeopardy if change was not forthcoming.

The vast clinical experience gathered by Federal surgeons filled *The Medical and Surgical History of the War of the Rebellion*, the official medical history of the US Civil War. The surgeons' experience with anesthetics, effective antimalarial medications, trauma surgery, open-air recovery, and sanitation marked a path toward better medicine in the Army, nation, and the world. Today's effective practice of clearing the battlefield follows the principles laid out in Dr Letterman's plan. The task of the horse-drawn wagon ambulance is now performed by the horsepowered rotary aircraft.

> **Scott C. Woodard**
> Historian
> US Army Medical Department
> Center of History and Heritage

Wounds of War

The typical American Civil War soldier faced a 25% probability that he would not survive the conflict. The odds were also two out of three that the cause of death would be disease and not combat. Dysentery and diarrhea killed far more men than bullets.

Small arms fire caused 94% of all wounds, with artillery accounting for the rest. More than 70% of all wounds were to the arms or legs. Penetrating wounds to the skull killed 80% of those so wounded, while wounds to the abdomen and chest claimed approximately 74% for each. Spinal wounds killed 56% of the soldiers. In all, the Union forces suffered approximately 110,000 battle deaths, while the Confederates lost about 94,000. Diseases claimed approximately 416,000 Northern and Southern soldiers during the 4 years of war.

Field dressings were not issued to soldiers in battle, and they received no first-aid training. Those who survived their initial treatment at a regimental aid station usually faced a 3 to 5 mile ambulance ride to reach a major field hospital.

Inexperience was the most common trait shared among US Civil War physicians. Few were properly trained in military medicine. Out of approximately 11,000 doctors in the Union Army, only about 500 had received any prewar surgical training. Out of about 3,000 Confederate Army doctors, a mere 27 were school-trained surgeons. Most doctors on both sides had never treated a gunshot wound prior to the war's start in 1861.

The most common form of surgery was amputation. Fully 75% of all wartime surgical procedures involved the removal of a mangled or infected limb. The overall mortality rate for amputations was 26%. Amputations of the leg at the hip joint killed the most soldiers (85%), while the loss of a foot at the ankle resulted in the deaths of only 13% of all patients. Shock and blood loss claimed many, but postoperative infections, particularly gangrene, were the greatest threat to survival in the wake of amputations.[2]

> Wayne R. Austerman, PhD
> Historian
> Center of Leader Training (Retired)

Orientation: Visitor Center

The Visitor Center at the Antietam National Battlefield in Sharpsburg, Maryland, provides a wonderful introduction to the Battle of Antietam and the Maryland Campaign. Within the center's museum, there are several medical-related exhibits and, in particular, objects from Assistant Surgeon William B. Wheeler of the Union 8th Maryland Infantry. His unit was assigned to Kenly's Maryland Brigade (Brigadier General John R. Kenly) in VIII Corps and arrived at Antietam on 18 September 1862. His V Corps insignia reflects his last unit of assignment.

Utensil Display Case. Medical instruments displayed at the Antietam National Battlefield's Visitor Center in Sharpsburg, Maryland, were used by Assistant Surgeon William B. Wheeler, of the Union 8th Maryland Infantry. Mary Mitchell, a witness to the battle's devastation just across the Potomac River in Shepherdstown, West Virginia, wrote in 1886 that the wounded "filled every building and overflowed into the country round, into farm houses, barns, corn cribs, cabins—wherever four walls and a roof were found together." The Union Army established more than 70 field hospitals in the area, which cared for their men and the Confederates who had been left behind. Courtesy of Scott C. Woodard, US Army Medical Department Center of History and Heritage; November 2017.

Inside the center's 134-seat theater, the award winning film, "Antietam Visit," is normally shown every half-hour. This 26-minute movie is an overview of the Battle of Antietam and tells about President Abraham Lincoln's visit with Union Commander General George B. McClellan. "Antietam," a one-hour documentary, is shown daily at noon. This film, narrated by James Earl Jones and filmed on the battlefield, follows General Robert E. Lee's entire Maryland Campaign.

Battlefield Orientation Talks are also offered every day. During the summer season, more scheduled talks, walks, and tours are conducted by National Park Rangers. The Visitor Center offers a daily schedule. Visitor Center hours are 9 am to 5 pm every day (except Thanksgiving, Christmas, and New Year's Day). The park entrance fee is $4 per adult; however, military groups who visit the park as part of their professional military education can qualify for a waiver of the entrance fee. Group entrance fees can be waived by submitting a fee waiver request. Visit the Antietam National Battlefield website (www.nps.gov/anti) for more information.

Dr Wheeler Display Case. A sack coat and compass belonging to Assistant Surgeon William B. Wheeler, of the Union 8th Maryland Infantry, are displayed at the Antietam National Battlefield's Visitor Center in Sharpsburg, Maryland. Courtesy of Scott C. Woodard, US Army Medical Department Center of History and Heritage; November 2017.

Inherent in all outdoor field activities is the need for participants to take a deliberate risk assessment of the weather and terrain during their investigation of the battlefield. Carrying plenty of water, applying sunscreen, and utilizing proper footwear and clothing will help mitigate potential risks.

Nearby Hospitals

NAME	ADDRESS	PHONE	DRIVE TIME (approx. minutes)	TRAUMA CENTER	HELIPAD	BURN CENTER
Meritus Medical Center	Hagerstown, MD	301-790-8000	25	Yes	Yes	
City Hospital	Martinsburg, WV	304-264-1000	30	Yes	Yes	
Martinsburg VA Medical Center	Martinsburg, WV	304-263-0811	25			
Jefferson Medical Center	Ranson, WV	304-728-1600	30			
The Johns Hopkins Hospital	Baltimore, MD	410-955-5000	more than 90	Yes	Yes	Yes

Hospital Locations. The information in this matrix is used by the Antietam National Battlefield staff when determining emergency care locations for real-world emergencies. Data courtesy of Olivia Black, volunteer coordinator at the Antietam National Battlefield in Sharpsburg, Maryland.

Recommended Sources

- Ballard T. *Staff Ride Guide: Battle of Antietam*. Washington, DC: US Army Center of Military History; 2008.
- American Battlefield Trust website. Accessed April 20, 2021. https://www.battlefields.org/ Note: Participants can search for the Antietam Battle App on the website; the application serves as a guide to all the historic spots on the 1862 Civil War battlefield.

REFERENCES

1. Robertson WG. *The Staff Ride*. Washington, DC: US Army Center of Military History; 1987:v.
2. Austerman WR. *Civil War Information Paper*. US Army Medical Department Center and School;1992.

Civil War Medical Structure

Civil War leaders demonstrated foresight when it came to providing medical support for soldiers in the field. Medical personnel were assigned to units from the regiment through the Army-level. Medical support above the regimental level was dependent upon the field commander's direction. Although there were some discrepancies, the general provisions for medical units and support were as follows:

Unit / Medical Support

Regiment, Infantry
Approx. 1000 soldiers

- 2-3 Surgeons (1 Surgeon was detached for hospital service support from the rear)
- 2 ambulances (3 men each) plus an NCO
- 1 wagon with supplies

Regiment, Cavalry
Approx. 800 soldiers

- 2 Surgeons
- 2 Farriers per troop (company)
- 1 Veterinary Sergeant
- 1 Veterinary Surgeon (after 24 March 1863)

Brigade, Infantry

- Surgeon-in-Chief (Brigade Surgeon)
- 3 ambulances (3 men each) plus Lieutenant
- 1 wagon with supplies

Division, Infantry

- Medical Director (Division Surgeon)
- 1 field hospital (3 operating teams)
- Ambulances under a Lieutenant

Corps

- Medical Director (Corps Surgeon)
- 1 Medical Inspector (staff officer)
- 1 Medical Officer in Charge of Hospitals
- Ambulances under a Captain

Army
eg. Army of the Potomac

- Medical Director (Army Surgeon)
- 1–2 Medical Inspectors (staff officers)
- 1 Medical Purveyor (supply officer)
- Ambulances under a Major

No Help from the War Department

Brigadier General (Dr) William A. Hammond was appointed as the new Surgeon General of the US Army on 25 April 1862, after the retirement of Brigadier General (Dr) Clement A. Finley. As a reformer, progressive medical practitioner, and administrator, Dr Hammond was the US Sanitary Commission's favored candidate. He had previously served as an assistant surgeon for 11 years, but resigned before American Civil War hostilities began to serve as a professor at the University of Maryland. Of course, coming back as the head of the US Army Medical Department, functionally skipping over more seasoned officers, did not go unnoticed by his political rivals.[1]

Dr Hammond had previously written to the US War Department citing the horrible Federal evacuation procedures. This plea was dismissed outright. It would later take a certain staff officer under an Army commander to properly influence battlefield evacuations.

Chap1RA **Portrait of Hammond.** Union Brigadier General (Dr) William A. Hammond, Surgeon General of the US Army from 1862–1864, led medical reform efforts in the Army and advocated for energetic and fellow reform-minded officers like Major (Dr) Jonathan Letterman. Source: Library of Congress. Brady, William; circa 1860–1865.

The following correspondence is from *The Medical and Surgical History of the War of the Rebellion (1861-65), Part III, Volume II, Surgical History:*[2]

"SURGEON GENERAL'S OFFICE.
WASHINGTON CITY, D.C., September 7, 1862.

"Sir: I have the honor to ask your attention to the frightful state of disorder existing in the arrangements for removing the wounded from the field of battle. The scarcity of ambulances, the want of organization, the drunkenness and incompetency of the drivers, the total absence of ambulance attendants, are now working their legitimate results, results which I feel I have no right to keep from the knowledge of the Department. The whole system should be under the charge of the Medical Department; an ambulance corps should be organized and set in instant operation. I have already laid before you a plan for such an organization, which I think covers the whole ground, but which I am sorry to find does not meet with the approval of the General-in-Chief. I am not wedded to it. I only ask that some system may be adopted by which the removal of the sick from the field of battle may be speedily accomplished, and the suffering to which they are now subjected be, in future, as far as possible avoided. Up to this date six hundred wounded still remain on the battle-field consequence of an insufficiency of ambulances and a want of a proper system for regulating their removal in the Army of Virginia. Many have died of starvation, many more will die in consequence of exhaustion, and all have endured torments which might have been avoided. I ask, sir that you will give me your aid in this matter; that you will interpose to prevent a recurrence of such consequences as have followed the recent battle, consequences which will inevitably ensue on the next important engagement, if something is not done to obviate them.

"I am, sir, very respectfully,
"Your obedient servant,
"WILLIAM A. HAMMOND,
"Surgeon General, U.S.A.

Doctor Jonathan Letterman, 1824–1872

Chap1RA **Major Letterman.** Major (Dr) Jonathan Letterman, Medical Director of the Army of the Potomac, spearheaded concrete medical readiness reforms that were eventually replicated throughout the entire US Army during the US Civil War. Source: US National Library of Medicine; circa 1862.

Jonathan Letterman was born in Canonsburg, Pennsylvania, on 11 December 1824, and graduated from Philadelphia's Jefferson College in 1845. Following an apprenticeship with his physician father, he graduated from Jefferson Medical College in 1849. Coincidentally, the medical school was originally established by his future commander's father, Dr George McClellan. Dr Letterman competed for a commission in the US Army Medical Corps along with 51 other applicants in an examination board held in the state of New York. He was one of nine who was offered a commission. It was here that Dr Letterman met Dr Hammond, another successful candidate in that intense review board.

Assistant Surgeon Letterman's early assignments included Fort Meade, Florida (1849–1853); Fort Ripley, Minnesota (1853–1854); Fort Leavenworth, Kansas (1854); Fort Defiance, New Mexico Territory (1854); Fort Union, New Mexico (1855–1858); Fort Monroe, Virginia (1858–1859); New York City (1859); and the Department of California (1860–1861).

During this time in the US Army, medical officers' ranks were actually descriptive of their positions. For example, doctors serving with the rank of "assistant surgeon," which was equivalent to a lieutenant or captain, could serve in staff positions and inspect hospitals within a command. The rank of "surgeon" was equivalent to a major.

Chap1RA **Soldiers and Abraham Lincoln.**
President Abraham Lincoln visits with Union Major General George B. McClellan and a group of officers from the Army of the Potomac near the Antietam battlefield in October 1862. Many famous and soon-to-be famous officers are in the group. They are (from left to right): 1) Colonel Delos B. Sacket, Inspector General; 2) Captain George Monteith; 3) Lieutenant Colonel Nelson B. Sweitzer; 4) General George W. Morell; 5) Colonel Alexander S. Webb, Chief of Staff, US V Corps; 6) General George B. McClellan; 7) Scout Adams; 8) Dr Jonathan Letterman, Army Medical Director; 9) Unknown; 10) President Abraham Lincoln; 11) General Henry J. Hunt; 12) General Fitz-John Porter; 13) Unknown; 14) Colonel Frederick T. Locke, Assistant Adjutant General; 15) General Andrew A. Humphreys; 16) Captain George Armstrong Custer. Source: Library of Congress. Gardner, Alexander; 1862.

In those grooming positions, Dr Letterman performed the expected duties of a frontier Army surgeon. He conducted inspections to ensure conditions were sanitary, treated the sick and wounded, reported meteorological conditions, and campaigned against Apache and Paiute Indians.[3]

But it was in the early portion of the War of the Rebellion that serendipity played its hand. As the Medical Director of West Virginia, Dr Letterman knew Major General George B. McClellan, Commander of the Department of the Ohio; he also knew Hospital Inspector William A. Hammond. McClellan was eager to reform the Army and was an early champion of the US Sanitary Commission. The US Sanitary Commission was a nongovernmental agency that advocated for Army medical reform and, functionally, filled the gap in Army treatment and evacuation. This same organization, in turn, advocated for the ascendancy of Dr Hammond to the position of Surgeon General. By the time McClellan assumed command of the Army of the Potomac in July 1861, and after subsequent medical disasters on the Virginia Peninsula, the Army was ripe for a change in medical leadership. Unfortunately, the initiatives of then Major (Dr) Charles S. Tripler, previous Medical Director of the Army of the Potomac (Dr Letterman's predecessor), were not implemented by the Army. In addition, the Army's Medical Department was considered by many to be disorganized and poorly run.[3]

Dr Letterman, who was then an Army major, assumed medical directorship on 4 July 1862 of a unified army of 100,000 soldiers under McClellan's Army of the Potomac. The Army's entire movement from the Virginia Peninsula to western Maryland was an exercise in logistics, as ambulances and equipment were added en route.[4] Dr Letterman's appearance swept in a wave of reform. The new director wrote the following:

> The subject of the ambulances became, after the health of the troops, a matter of importance. No system had anywhere been devised for their management. They were under the control both of Medical officers and Quartermasters, and, as a natural consequence, little care was exercised over them by either. They could not be depended upon for efficient service in time of action or upon a march, and were too often used as if they had been made for the convenience of commanding officers. The system I devised was based upon the idea that they should not be under the immediate control of Medical officers, whose duties, especially on the day of battle, would prevent any proper supervision; but that other officers, appointed for that especial purpose, should have direct charge of the horses, harness, ambulances, etc., and yet under such regulations as would enable Medical officers at all times to procure them with facility when needed for their legitimate purpose.[5]

Dr Letterman's reform movement was based on his keen sense of what was needed to improve medical readiness and on the importance of relationship building. He saw beyond myopic circumstances and tactical scenarios. His vision was far-reaching and bold.

The Army's inability to evacuate patients and push them to surgical care was an indicator of operational failures. Dr Letterman saw the bigger problem and understood his purpose in the Army. He knew the readiness of the Army to win in combat was measured by its ability to employ the skills of its primary weapon—the soldier. The sick and dead cannot fight.

Following the war, Dr Letterman expressed these thoughts:

> A corps of Medical officers was not established solely for the purpose of attending the wounded and sick; the proper treatment of these sufferers is certainly a matter of very great importance, and is an imperative duty, but the labors of Medical officers cover a more extended field. The leading idea, which should be constantly kept in view, is to strengthen the hands of the Commanding General by keeping his army in the most vigorous health, thus rendering it, in the highest degree, efficient for enduring fatigue and privation, and for fighting.[6]

Establishment of the Ambulance Corps, Part I

Chap1RA **Dr Letterman and Staff.**
Major (Dr) Jonathan Letterman, Medical Director of the Army of the Potomac (seated in front of the pole), with his staff at Warrenton, Virginia. Behind him (left to right) are two contract surgeons, a medical cadet, assistant surgeon, and orderly. Source: Library of Congress. Gardner, Alexander; 1862.

The Battle of Antietam was the single bloodiest battle in the history of the US Army and was fought outside of Sharpsburg, Maryland. It was Major (Dr) Jonathan Letterman's first engagement as the Medical Director of the Army of the Potomac. His plan to reorganize the ambulance system was issued on 2 August 1862, but it was not fully implemented before the day of the battle on 17 September 1862.

In his correspondence with the Office of the Surgeon General, Dr Letterman submitted the following observations in his report on the Medical Department of the Army of the Potomac from 4 July to 31 December 1862.

> ... The subject of the ambulances, after the health of the troops, became a matter of importance. Medical officers and quartermasters had charge of them, and, as a natural consequence, little care was exercised over them, and they could not be depended upon during an action or upon a march. It became necessary to institute some system for their management, such that they should not be under the immediate control of medical officers, whose duties, especially on the day of battle, prevented any supervision, when supervision was, more than at any other time, required. It seemed to me necessary, that whilst medical officers should not have the care of the horses, harness, etc., belonging to the ambulances, the system should be such as to enable them, at all times, to procure them with facility when wanted for the purpose for which they were designed, and to be kept under the general control of the medical department. Neither the kind nor the number of ambulances required were in the army at that time, but it nevertheless was necessary to devise a system that would render as available as possible the material upon the spot, particularly as the army might move at any time, and it was not considered advisable to wait for the arrival of such as had been asked for, only a portion of which ever came. In order to inaugurate a system which would make the best of the materials on hand and accomplish the objects just referred to, the following order was written and published by direction of the commanding general...."[7]

Establishment of the Ambulance Corps, Part II

Although it would be years before the entire US Army developed the Ambulance Corps that Dr Letterman envisioned, Dr Letterman proved to be a highly effective staff officer. General Orders No. 147 was issued "by command of Major General McClellan," but the work of Surgeon (Dr) Letterman was evident. It was considered the first official establishment of an ambulance corps in the US Army. The order was all-encompassing in scope and directed responsibilities at each level of command. It established the

Chap1RA **Side and Rear View of the Finley.**
A side and rear view of the Finley two-wheeled (one horse) ambulance. Most regiments had these two-wheeled ambulances; however, their light construction proved to be too frail for rough country roads. Washington, DC: US Government Printing Office; 1870.

Chap1RA **Side View of the Wheeling.** Side view of the Wheeling-Rosecrans ambulance wagon. These rugged ambulances fulfilled their purpose of safely carrying patients and hauling medical supplies. Washington, DC: US Government Printing Office; 1870.

10 Combat Readiness Through Medicine at the Battle of Antietam

requirement to inspect, drill, maintain, and supervise. The order also treated the ambulance as a "system" where the support materiel and equipment were vital to the end state of patient care.

General Orders No. 147 provided an official definition for the use of ambulance wagons and, most importantly, for what was prohibited. Most unique, in addition to the specially designated headgear and sleeve insignia, was the authorization for US Army Medical personnel in the Ambulance Corps to carry a revolver—unlike the surgeon's sword which was intended for show only.

"HEADQUARTERS ARMY OF THE POTOMAC,
CAMP NEAR HARRISON'S LANDINGS, VA., *August* 2, 1862.
"GENERAL ORDERS No. 147.

The following regulations for the organization of the Ambulance Corps and the management of ambulance trains are published for the information and government of all concerned. Commanders of Army Corps will see that they are carried into effect without delay:

1. The Ambulance Corps will be organized on the basis of a captain to each Army Corps as the Commandant of the Ambulance Corps, a 1st lieutenant for a Division, 2d lieutenant for a brigade, and a sergeant for each regiment.
2. The allowance of ambulances and transport carts will be: One transport cart, one 4-horse and two 2-horse ambulances for a regiment; one 2-horse ambulance for each battery of artillery, and two 2-horse ambulances for the Headquarters of each Army Corps. Each ambulance will be provided with two stretchers.
3. The privates of the Ambulance Corps will consist of two men and a driver to each ambulance, and one driver to each transport cart.
4. The captain is the commander of all the ambulances and transport carts in the Army Corps, under the direction of the Medical Director. He will pay special attention to the condition of the ambulances, horses, harness, etc., requiring daily inspections to be made by the commanders of Division ambulances, and reports thereof to be made to him by these officers. He will make a personal inspection once a week of all the ambulances, transport carts, horses, harness, etc., whether they have been used for any other purpose than the transportation of the sick or wounded, and medical supplies; reports of which will be transmitted through the Medical Director of the Army Corps to the Medical Director of the Army every Sunday morning. He will institute a drill in his corps, instructing his men in the most easy and expeditious method of putting men in and taking them out of the ambulances, taking men from the ground and placing and carrying them on stretchers, observing that the front man steps off with the left foot and the rear man with the right, etc. He will be especially careful that the ambulances and transport carts are at all times in order, provided with attendants, drivers, horses, etc., and the keg daily rinsed and filled

with fresh water, that he may be able to move at any moment. Previous to and in time of action he will receive from the Medical Director of the Army Corps his orders for the distribution of the ambulances and the points to which he will carry the wounded, using the light two horse-ambulances for bringing men from the field, and the four-horse ones for carrying those already attended to farther to the rear, if the Medical Director considers it necessary. He will give his personal attention to the removal of the sick and wounded from the field and to and from the hospitals, going from point to point to ascertain what may be wanted, and to see that his subordinates (for whose conduct he will be responsible) attend to their duties in taking care of the wounded, treating them with gentleness and care, and removing them as quickly as possible to the places pointed out; and that the ambulances reach their destination. He will make a full and detailed report after every action and march of the operations of the Ambulance Corps.

5. The 1st lieutenant assigned to the Ambulance Corps of a Division will have complete control, under the Commander of the whole Corps and the Medical Director, of all the ambulances, transport carts, ambulance horses, etc., in the Division. He will be the Acting Assistant Quartermaster for the Division Ambulance Corps, and will receipt and be responsible for the property belonging to it, and be held responsible for any deficiencies in ambulances, transport carts, horses, harness, etc., pertaining to the Ambulance Corps of the Division. He will have a travelling cavalry forge, a blacksmith, and a saddler, who will be under his orders, to enable him to keep his train in order. He will receive a daily inspection report of all the ambulances, horses, etc., under his charge from the officer in charge of Brigade Ambulance Corps, will see that the subordinates attend strictly to their duties at all times, and will inspect the corps under his charge once a week; a report of which inspection he will transmit to the Commander of the Ambulance Corps.

6. The 2d lieutenant in command of the ambulances of a brigade will be under the immediate orders of the commander of the Ambulance Corps for the Division, and have superintendence of the Ambulance Corps for the brigade.

7. The sergeant in charge of the Ambulance Corps for a regiment will conduct the drills, inspections, etc., under the orders of the Commander of the Brigade Ambulance Corps, and will be particular in enforcing rigidly all orders he may receive from his superior officers. The officers and non-commissioned officers of this corps will be mounted.

8. The detail for this corps will be made with care by Commanders of Army Corps, and no officer or man will be selected for this duty except those known to be active and efficient; and no man will be relieved except by orders from these Headquarters. Should any officer or man detailed for this duty be found not fitted for it, representations of the fact will be made by the Medical Director of the Army Corps to the Medical Director of this Army.

9. Two medical officers from the reserve corps of surgeons of each Division, and a hospital steward who will be with the medicine wagon, will be detailed by the Medical Director of the Army Corps to accompany the ambulance train when on the march, the train of each Division being kept together, and will see that the sick and wounded are properly attended to. A medicine wagon will accompany each train.

10. The officers connected with the corps must be with the trains on the march, observing that no one rides in the ambulances without the authority of the medical officers, except in urgent cases; but men must not be allowed to suffer, and the officers will, when the medical officers cannot be found, use a sound discretion in this matter, and be especially careful that the men and drivers are in their proper places. The place for the ambulances is in front of all the wagon trains.

11. When in camp the ambulances, transport carts, and Ambulance Corps will be parked with the brigade, under the supervision of the commander of the corps for the brigade. They will be used on the requisition of the regimental medical officers, transmitted to the Commander of the Brigade Ambulance Corps, for transporting the sick to various points and procuring medical supplies, and for nothing else. The non-commissioned officer in charge will always accompany the ambulances or transport carts when on this or any other duty, and he will be held responsible that they are used for none other than their legitimate purposes. Should any officer infringe upon this order regarding the uses of ambulances, etc., he will be reported by the officer or non-commissioned officer in charge to the commander of the train, all the particulars being given.

12. The officer in charge of a train will at once remove anything not legitimate, and if there be not room for it in the baggage wagons of the regiment, will leave it on the road. Any attempt by a superior officer to prevent him from doing his duty in this or any other instance he will promptly report to the Medical Director of the Army Corps, who will lay the matter before the commander of that corps. The latter will, at the earliest possible moment, place the officer offending in arrest for trial for disobedience of orders.

13. Good, serviceable horses will be used for the ambulances and transport carts, and will not be taken for any other purpose, except by orders from these Headquarters.

14. The uniform of this corps is: For privates, a green band two inches broad around the cap, a green half chevron two inches broad on each arm above the elbow, and to be armed with revolvers. Non-commissioned officers to wear the same band around the cap as a private, chevrons two inches broad, and green, with the point toward the shoulder, on each arm above the elbow.

15. No person will be allowed to carry from the field any wounded or sick, except this corps.

16. The Commanders of the Ambulance Corps, on being detailed, will report without delay to the Medical Director at these Headquarters for instructions. All Division, Brigade, or Regimental Quartermasters having any ambulances, transport carts, ambulance horses or harness, etc., in their possession, will turn them in at once to the Commander of the Division Ambulance Corps.

BY COMMAND OF MAJOR GENERAL McCLELLAN:
(Signed) S. WILLIAMS,
Assistant Adjutant-General[8]

As recorded in the official record, Dr Letterman's efforts greatly enhanced medical readiness in the Army of the Potomac.

Chap1RA Ambulance Drill.
Newly organized Ambulance Corps members rehearse their battle drills soon after the Battle of Antietam. Source: Library of Congress; 1862.

Dr Letterman wrote the following:

> The advantages accruing from this organization became speedily manifest. At the battle of Antietam, in September, 1862, by the active and energetic exertions of the members of this corps the disabled of the right wing of the army (there was no ambulance system on the left) were rapidly conveyed from the scene of conflict to the hospitals in the rear. The train of ambulances plied incessantly between the battleground and the field hospital. During the night of the battle, all of our wounded in the widely

extended field were removed to shelter and received the necessary surgical attention. Different members of the corps behaved with the utmost gallantry, passing freely under fire in their search of the fallen, and advancing at times to the extreme verge of the enemy's pickets. All of our wounded having thus been collected at the temporary depots, such as were deemed best able to undergo further transportation were carefully selected. These, during the following two or three days, were then conveyed by the ambulance train to Frederick City, Md., the nearest point of railway connections. At the first battle of Fredericksburg the results of the persevering endeavors of the Ambulance Corps were not less happy. During the night following the battle all of the wounded remaining on the ground not absolutely in the hands of the enemy were safely conveyed to the city of Fredericksburg and its neighborhood. When it subsequently became necessary to evacuate the city of Fredericksburg for military reasons, the wounded were again placed upon the train and safely reached the opposite bank of the river. These fortunate results were, however, obtained at the expense of the Ambulance Corps, which experienced a loss of one officer and several privates killed, besides others who were captured during their humane efforts to remove their fallen countrymen."[9]

Health of the Confederate Commander

General Robert E. Lee, 55 years old at the time of the battle, suffered a slight wound in the Mexican–American War and, in 1849, contracted malaria in Baltimore, Maryland. On 31 August 1862, he suffered a fall, which left one hand badly sprained and the other with several broken bones. Both hands were placed in splints, with one hand in a sling. He was unable to ride a horse for 10 days. At the Battle of Antietam, the injuries were at worst an irritant.[10]

Chap1RA **General Robert E. Lee.**
Confederate General Robert Edward Lee, Commander of the Army of Northern Virginia, poses for a full-length portrait wearing his dress uniform and sword. Source: Library of Congress. Vannerson, Julian; circa 1861–1865.

Union Commander's Health Profile

Major General George B. McClellan was 36 years old at the time of Antietam. In August 1847, he suffered a bruising injury in the Mexican–American War and was hospitalized the following year for over a month with dysentery and malaria. Both of these conditions recurred periodically for the rest of his life. In 1853, he was prescribed laudanum for chills, fever, and jaw pain. In the summer of 1861, he was stricken with malaria and typhoid fever, rendering him ill for a month. In May 1862, he had recurrences of dysentery for a week, followed by 10 days of physical exhaustion. He was stricken with diarrhea on 27 August and again following the Battle of Antietam. On 7 November 1862, McClellan was relieved of command.[11]

Chap1RA **Major General McClellan.** Union Major General George Brinton McClellan, Commander of the Army of the Potomac, poses for a full-length portrait with hand in coat, a standard of the day. Source: Library of Congress. Brady, William; circa 1861–1865.

CHAPTER 1 READ AHEAD REFERENCES

1. Gillett MC. *The Army Medical Department, 1818-1865*. Washington, DC: Center of Military History, US Army; 2000:177–179.
2. Otis GA. *The Medical and Surgical History of the War of the Rebellion (1861-65), Part III, Volume II, Surgical History*. Washington, DC: US Government Printing Office; 1883:934.
3. Marble S (ed.). *Builders of Trust: Biographical Profiles from the Medical Corps Coin*. Fort Detrick, MD: The Borden Institute; 2011:35–39.
4. Ryons FB. The United States Army Medical Department 1861-1865. *Milit Surg*. 1936:353–355.
5. Letterman J. *Medical Recollections of the Army of the Potomac*. New York, NY: D. Appleton & Co; 1866:22–23.
6. Letterman J. *Medical Recollections of the Army of the Potomac*. New York, NY: D. Appleton & Co; 1866:100.
7. Woodward JJ. *The Medical and Surgical History of the War of the Rebellion (1861-65), Part I, Volume I, Medical History, Appendix to Part I, Containing Reports of Medical Directors, and Other Documents*. Washington, DC: US Government Printing Office; 1870:95–96.
8. Letterman J. *Medical Recollections of the Army of the Potomac*. New York, NY: D. Appleton & Co; 1866:24–30.
9. Otis GA. *The Medical and Surgical History of the War of the Rebellion (1861-65), Part III, Volume II, Surgical History*. Washington, DC: US Government Printing Office; 1883:935–938.
10. Welsh JD. *Medical Histories of Confederate Generals*. Kent, OH, and London, England: The Kent State University Press; 1995:134.
11. Welsh JD. *Medical Histories of Union Generals*. Kent, OH, and London, England: The Kent State University Press; 1996:937–938.

Medical Annex to the Battle of Antietam

This map will help participants orient their positions on the Antietam battlefield as they read the descriptions in the narrative. Each stop along the way is annotated with a corresponding chapter and number and follows the progression of the battle. The original map is courtesy of the National Park Service and has been modified to reflect the key positions outlined in this book.

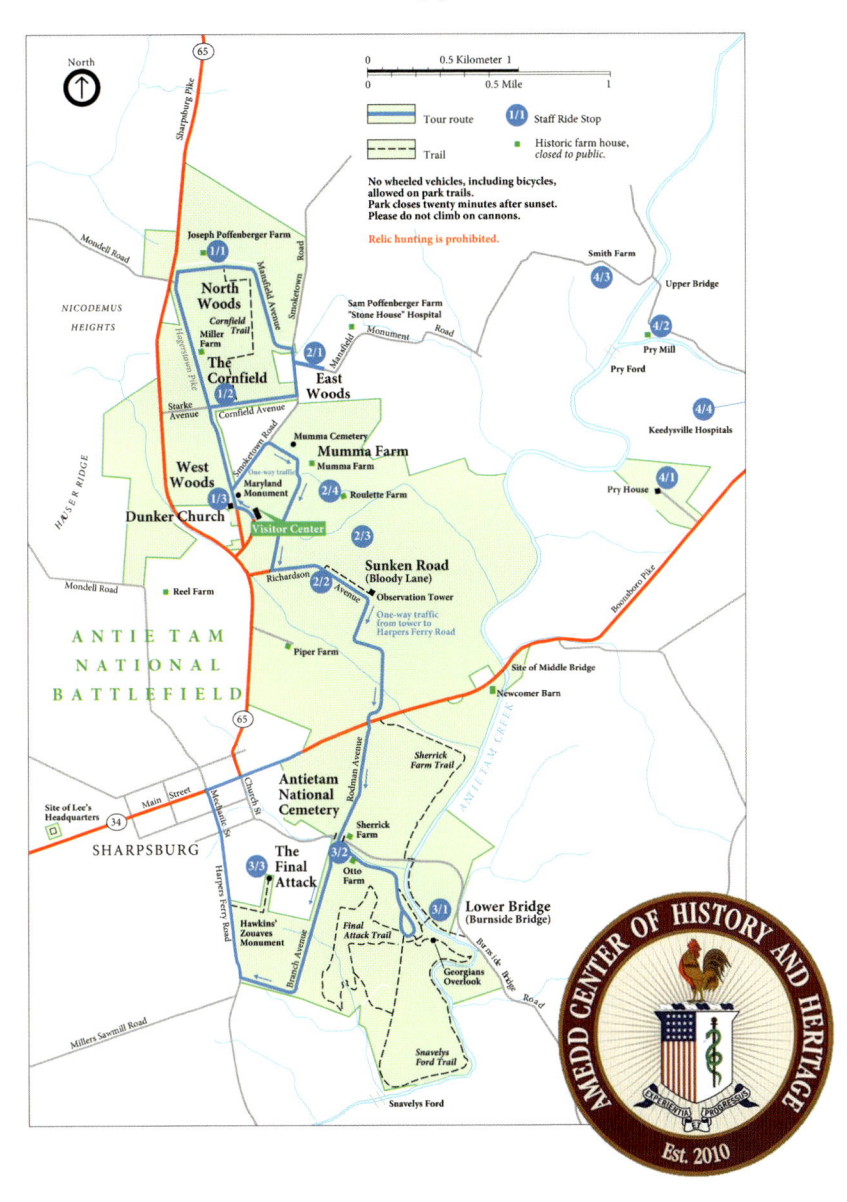

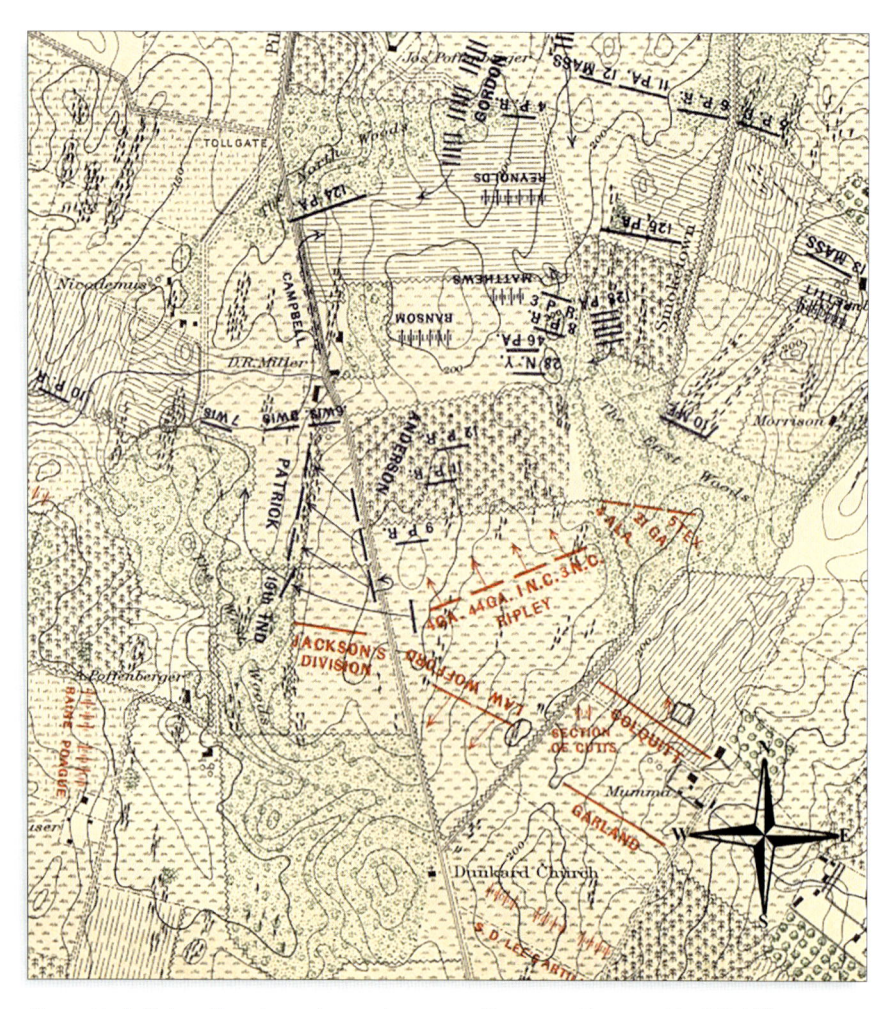

Figure 1.1 **Antietam Map.** An early morning array of troops gather near the D.R. Miller Cornfield on 17 September 1862. The Union 12th Massachusetts Infantry and the Union 124th Pennsylvania Infantry, described in the following pages, are indicated on the upper part of map. Source: US War Department. *Atlas of the battlefield of Antietam*. Washington, DC: US Government Printing Office; 1908.

CHAPTER 1

Direct Action on the Battlefield

1/1 STOP 1: NORTH WOODS

The foliage today in the North Woods at the Antietam National Battlefield is not as dense as it was on the morning of 17 September 1862. Facing south, it was here in these woods (**Chapter 1, Stop 1**) in Sharpsburg, Maryland, that the Union troops assembled for the upcoming battle. Within the wooded canopy, regimental hospitals awaited their gruesome collections in makeshift shelters or in buildings of opportunity on local farms. Nearby, the bulk of Hooker's I Corps (Union Major General Joseph Hooker) and Mansfield's XII Corps (Union General Joseph K. Mansfield), both part of the Army of the Potomac, were staging for the imminent duel. Jackson's Corps (Confederate Major General Thomas J. Jackson), of the Army of Northern Virginia, was positioned with artillery looking north from its location near the Dunker Church. The troops of Hood's Division (Confederate Brigadier General John B. Hood) stood ready to reinforce the line. Along the ridge to the left (as staff ride participants face the church), the Southern "King of Battle" (nickname for artillery) was positioned to unleash what witnesses described as "Artillery Hell."

The field of corn to the left of the D.R. Miller Farm was contested all morning with Union and Confederate troops trading off possession.

Schell Amongst the Shells

Figure 1.2 **Hospital Illustration.** The field hospital located near the North Woods on the morning of 17 September 1862 as depicted by illustrator Frank H. Schell. The scene details a bloody amputation and shows limbs from other amputations scattered in the background. Source: Heritage Auctions, HA.com.

Frank H. Schell, the famous illustrator of *Frank Leslie's Illustrated Newspaper*, described the hospital scenes before him as he traveled through the battlefield near the North and East Woods that morning.

> Beside a fringe of timber was a farm already doing full duty as a field hospital, with a strangely pretty artistic effect of tents and tiny arbors grouped about its various buildings. Hastily noting the distressing work of the surgeon's knife and saw, I hurried away from its rueful accompaniment of shrieks and groans. . . . [A] shell exploded with terrifying report over the farm hospital I had sketched, sending its hissing fragments among the wounded and

attendants, and causing such anxious commotion that I hastened back. Preparations were being made to locate it further to the rear. Two-wheeled ambulances, coming up with their groaning passengers, were ordered "out the Smoketown road," to the farm determined on for the removal. . . . Numbers of brave stretcher bearers and some ambulances, jostling and jerking their tortured charges over the rough ground, passed rearward from the line of the wood, or were gleaning the field of the few helplessly wounded remaining among the scattered corpses here and there to be seen amid the long grass.[1]

Moving along the battlefield, Schell could see the Roulette Farm (**Chapter 2, Stop 4**) to his south from the East Woods.

> Ambulances were coming and taking available stations in the wood, where they had not long to wait for business. The fields began again to swarm with the litter and stretcher men, and with wounded soldiers who, walking with or without assistance, were wearily plodding back in search of the field hospitals.[2]

Closer to him on his right, as he peered toward the center of the battlefield, Schell could see the Mumma Farm afire.

> . . . I came to the smoldering ruins of Mummer's [Mumma] Farm buildings, where stretcher-bearers, resting their agonized burdens and their tired selves in the fancied security of the smoke-screened spot, were awaiting the ambulances. "Canister!" someone shouted, and, with a common impulse, all who could jumped for the protection of the stone wall of the burned barn, against which, with a sound as of heavy hailstones against window glass, deadly missiles rattled harmlessly. "What, boys! Your helpless comrades out there in the hellstorm?" It was but a moment of nerve shock, and the wounded were heroically brought in under shelter; one with a fresh "plug," that left him a horribly mangled corpse upon a bloodsopped litter. His would-be rescuer, wounded in the attempt, taking his place upon the freshly bespattered canvas.[3]

Direct Action on the Battlefield 23

The First of Three Wounds

Union Private Edson D. Bemis received the first of his three wounds during the American Civil War at the Battle of Antietam. Bemis and his fellow Company K soldiers, part of the 12th Massachusetts Infantry, assembled in the North Woods. The 12th Massachusetts, under Ricketts' Division (Union Major General James B. Rickett), along with Hooker's I Corps, charged into the eastern portion of D.R. Miller's Cornfield to their south. It was in this area that Bemis was wounded by a musket ball, which fractured the shaft of his upper left arm. According to known doctrine and tactics, the regimental hospital would have been located near the North Woods in an area protected by trees or a makeshift shelter. His arm eventually healed, but rested a quarter of an inch shorter than his right arm.

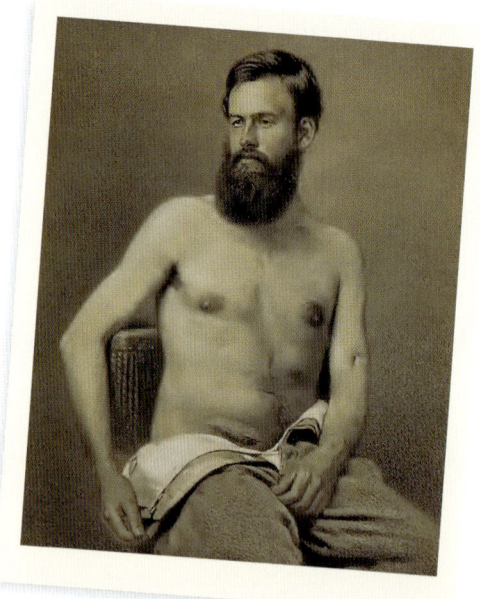

Figure 1.3 **Private Bemis Portrait.** Union Private Edson D. Bemis as shown in a drawing. Otis, George A. *Medical and Surgical History of the War of the Rebellion, Part I, Volume II, Surgical History*. Washington, DC: US Government Printing Office; 1870.

A detailed medical history of Private Edson D. Bemis can be found under "Selected Antietam Medical Histories" on Page 196.

Beating Long Odds on the Hagerstown Pike

With only three weeks of training, Union Private George D. Miller, Company D, and the other members of the 124th Pennsylvania Infantry, staged in the North Woods. Marching from Rockville, Maryland, they bypassed the Battle at South Mountain, but the evidence of it was clearly there. A cart full of appendages, mostly legs severed above the knee, stood watch as the troops passed by. A bloody pool encircled in bones and pieces of flesh marked the spot of a surgeon's good day of work. Miller was horrified by the sight and feared the old "saw bones" instruments more than the enemy's tools of death.

A little after 8 am, the 124th Pennsylvania moved out of the North Woods. Seven of the infantry's companies advanced along the Hagerstown Pike and into the D.R. Miller Cornfield, while the remaining three companies advanced past the Miller Barn on the Pike's western side. Confederate rifle fire lashed them from the West Woods and Miller was struck by a bullet, which passed through his torso and severed his colon.

After evacuating rearward to a field hospital, Miller lay in agony with a type of wound that proved fatal in almost all cases. He suffered for 10 days with fecal matter oozing out through the wound. The pain continued until his condition began to improve in early November. He miraculously eluded peritonitis (inflammation of the peritoneum—a membrane lining the abdomen). Because of his wound, Miller was discharged. He later returned home to West Chester, Pennsylvania, where he married, fathered three children, and lived in continuous severe pain from his injury until his death at 80 years old in 1919.[4]

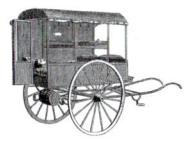

From the parking lot at the North Woods (Chapter 1, Stop 1), travel south to the lower end of the D.R. Miller Cornfield (Chapter 1, Stop 2).

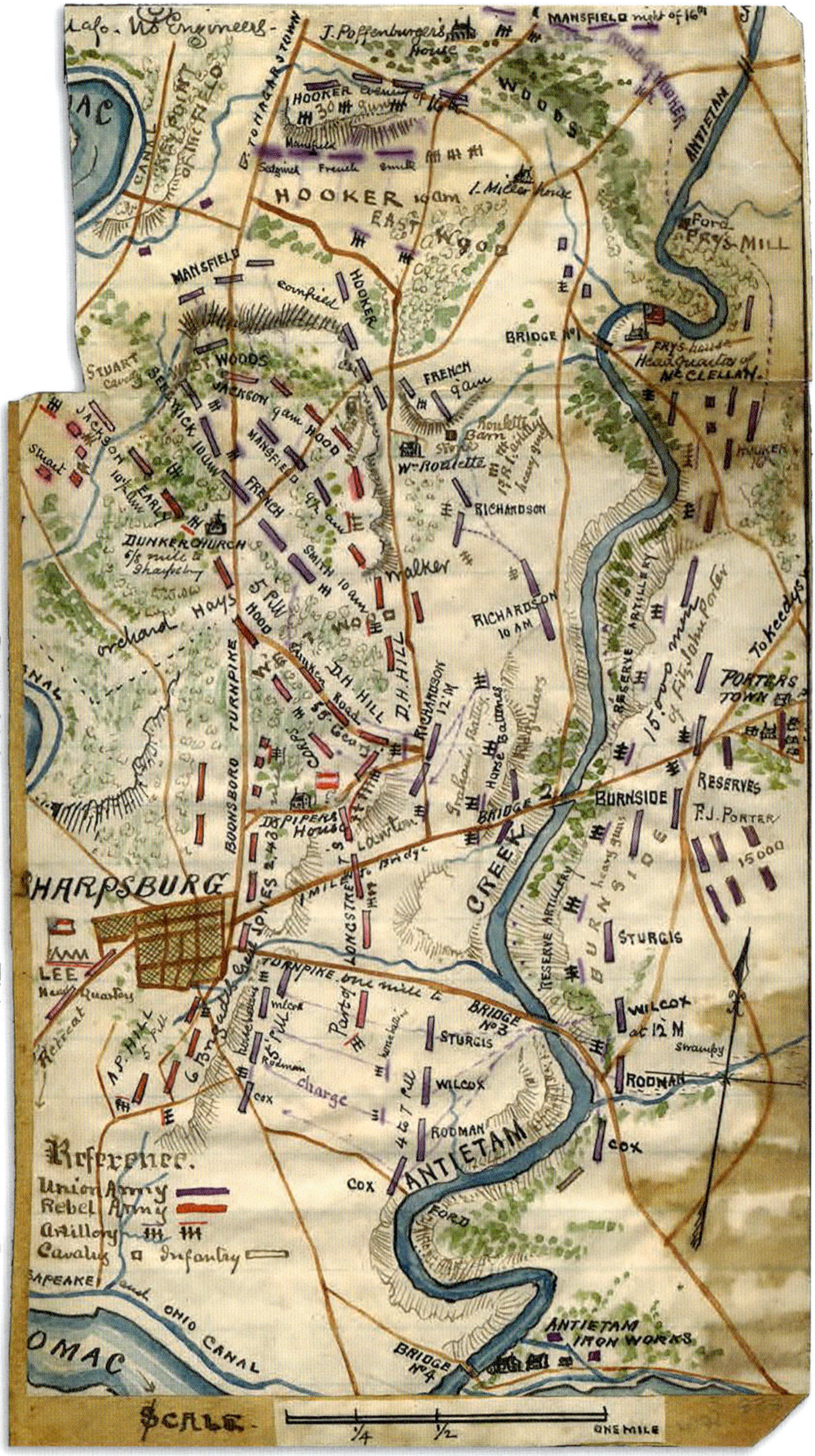

1/2 STOP 2: D.R. MILLER'S CORNFIELD

From this location (**Chapter 1, Stop 2**), looking north back toward the previous position, the D.R. Miller Cornfield today is a sterile reflection of its former self. The Miller Farm is halfway between the first stop on the left. The wounded from both sides of the war made their way to this makeshift hospital seeking succor and protection. In September of 1862, the corn would have been as tall as the average soldier and would have masked troop movements and exacerbated the struggle to retrieve the wounded. Federal cannons firing down from the front right and Confederate volleys from the back right collided in the bloody cornfield with the ferocity matched by clashing bayonets, bone splintering lead, and screams of both anger and fear. The ground was soaked with blood.

[Top]
Figure 1.4 **Modern View of D.R. Miller's Cornfield.**
In September of 1862, the corn would have been as tall as the average soldier and would have masked troop movements and exacerbated the struggle to retrieve the wounded. Source: National Park Service, US Department of the Interior.

[Opposite]
Figure 1.5 **Hand-Colored Map.**
This hand-colored map shows the forces (in the upper left portion) arranged around the D.R. Miller Cornfield on 17 September 1862. Sneden, Robert Knox. *Plan of the Battle of Antietam, Maryland*; circa 1862–1865.

A Surgeon's Effluvia

Union Surgeon Daniel M. Holt, 121st New York Infantry, wrote a letter to his wife, Mary Louisa Holt. In it, he described the grim realities facing him on the battlefield.

Figure 1.6 **View from Hagerstown Road.** The 1862 view from Hagerstown Road at the Battle of Antietam. The Confederate dead were left where they died. Later, they were eventually buried or retrieved by family members. Taking care of the dead was a low priority on the battlefield. The smells and horrors left their mark on the witnesses who were left alive. Source: Library of Congress. Gardner, Alexander; 1862.

Figure 1.7 **2017 View from Hagerstown Road.** The modern view from Hagerstown Road at the Antietam National Battlefield is in stark contrast to the putrid destruction of the battle. Courtesy of Scott C. Woodard, US Army Medical Department Center of History and Heritage; November 2017.

> I have seen, stretched along, in one straight line, ready for interment, at least a thousand blackened, bloated corpses with blood and gas protruding from every orifice, and maggots holding high carnival over their heads. Such sights, such smells and such repulsive feelings as overcome one, are with difficulty described. Then add the scores upon scores of dead horses, sometimes whole batteries lying alongside, still adding to the commingling mass of corruption and you get a faint, a very faint idea of what you see, and can always see after a sanguinary battle. Every house for miles around is a hospital and I have seen arms, legs, feet and hands lying in piles rotting in the blazing heat of a Southern sky unburied and uncared for, and still the knife went steadily in its work adding to the putrid mess.[5]

Lieutenant Stickley's Bloody Odyssey

Around 6:20 am, Confederate Lieutenant Ezra Stickley of Company A, 5th Virginia Infantry, ran out of luck. Federal artillery "walked in" three explosive rounds toward his butternut-clad brigade with each one landing successfully closer. The third ball killed Stickley's horse and wounded him as his unit waited for the enemy in the West Woods. He fell into the arms of terrified infantrymen. The shell ripped off his right arm, which dangled by shreds of skin at his shoulder. The shrapnel stripped the flesh from his right side; it also bruised a lung and broke a rib.

Stickley regained consciousness for an hour or so as he lay on the floor of an abandoned farmhouse (possibly at the Alfred Poffenberger Farm). Two surgeons prepared to tie off the arteries in the stump of his profusely bleeding right arm. However, they fled the scene before they finished amputating his arm when Union artillery began targeting the house. A soldier appeared and asked if Stickley needed help. Stickley asked for water, but there was none. The loyal soldier then pulled Stickley to his feet and began dragging him southeast toward the Hagerstown Pike.

Approximately 10 hours later, Stickley lay in the back of an ambulance that was evacuating south into Virginia. He was still semi-conscious after the amputation of his arm and the tying off of its arteries by doctors from Grigsby's Brigade (Confederate Colonel Andrew J. Grigsby). Stickley was woozy, but thankful to be alive after a trip on foot of several miles to find treatment.[6]

Figure 1.8 **Fight in the Cornfield.** Artist Arthur Lumley witnessed the Irish Brigade, which consisted mostly of Irish Americans who served in the Union Army, drive out the Rebels on the Federal right wing. Source: Library of Congress; Sharpsburg, MD; 1862.

Field Expedient Trauma Care

Gunner Ed Duffey of Parker's Virginia Battery (Confederate Captain William W. Parker) was pouring fire northward into D.R. Miller's Cornfield at about 7 am when a bullet ripped into his left thigh. Limping rearward, he found a dressing station where the surgeon offered him some pre-surgery whiskey. In behavior curious for an Irishman, he declined the whiskey in favor of water. The surgeon then cut the bullet out of his thigh without the benefit of an anesthetic.[7]

Sergeant Herzog's Discharge

Around 7:30 am, Battery B of the Union 4th US Artillery was attempting to hold its position just south of the D.R. Miller House at the northwest edge of the cornfield—as shattered Union regiments streamed past it to the rear. The Union gunners poured canister rounds into the oncoming soldiers of the Confederate 1st Texas Infantry, who replied with a devastating volley of musketry. Union Sergeant Joseph Herzog dropped gut shot (was hit in the bowels). Holding together the blue coils of his intestines in one hand, he stumbled northward for the dressing station in the Miller House.

At the Miller House, a surgeon grimly examined Herzog's wound and then advised the suffering man that he had at best a few hours left to live. "If that is the case, doctor," Herzog replied, "those few hours are not worth living." He then coolly drew his revolver and shot himself in the right temple.[8]

The Youngest Casualty

Around 1 pm, the Union 49th Pennsylvania Infantry arrived on the D.R. Miller Farm and took position on the right flank of the Union 1st Battery, New York Light Artillery. Sporadic Confederate artillery fire lashed the area. During one barrage, a shell burst among the ranks of the 49th Pennsylvania and wounded Company F drummer boy, 13-year-old Private Charles King. A piece of shrapnel hit "through the body." With great affection, members of his company evacuated the bleeding boy to a field hospital. King hung on in anguish until 20 September 1862. He was the youngest soldier on either side to die as a direct result of combat during the American Civil War.[9]

Self-Aid

Union Private Van R. Willard, of the 3rd Wisconsin Infantry, was part of the back-and-forth action across the D.R. Miller Cornfield. During the final push of the Confederates into the West Woods, Willard was wounded in the thigh.

> After being wounded, I limped away to one of the hospitals, but so great was the crowd of wounded already there that I thought I hobble on as far as I could. I went on until I came to a stream of water, where I drank, and filling my pail (a small tin one which I had carried for nearly half a year), I went into the woods and washed and bandaged my wound. Then fortunately, an ambulance came along. I got aboard and was taken to the church at Keedysville. It and all the other public buildings were converted into hospitals, as was also the case at Sharpsburg, Boonesboro, Middletown, and all the other towns between the battle field and Frederick City. I remained until Sunday the 21st, when I was taken to F. C. [Frederick City], where I remained until November the 10th. . . .[10]

Eventually, Willard was transferred by train to Philadelphia where he continued to recover from his extensive wound. He returned home to Wisconsin in 1864, went to law school, and became a successful businessman and civic leader. He passed away from cerebral apoplexia (stroke) in 1898.[11]

A Masonic Incident

The following is an excerpt from the *A History of the Fifth Regiment, New Hampshire Volunteers, in the American Civil War, 1861–1865*:

". . . The Fifth New Hampshire formed the picket line along the edge of the cornfield, where Richardson's Division [Major General Israel B. Richardson] fought. The reserve was in one edge of the corn, and the pickets about middle way of the field, concealed in the corn, as the sharpshooters of the enemy fired on all who undertook to walk around on the battlefield at that locality. Early in the morning one of the wounded rebels, who laid just outside the pickets, called one of the New Hampshire men and handed him a little slip of paper, on which he had, evidently with great difficulty, succeeded in making some mystic signs in a circle, with a bit of stick wet in blood. The soldier

was begged to hand the paper to some Freemason as soon as possible, and he took it to Colonel E. E. Cross of his regiment. The colonel was a Master Mason, but could not read the mystic token, it belonging to a higher degree. He, therefore, sent for Captain J. B. Perry, of the Fifth, who was a member of the 32nd degree of Freemasonry, and showed him the letter. Captain Perry at once said there was a brother Mason in great peril, and must be rescued. Colonel Cross instantly sent for several brother Masons in the regiment, told the story, and in a few moments four "brothers of the mystic tie" were crawling stealthily through the corn, to

Figure 1.9 **Surgeon William Child.** Surgeon William Child of the Union 5th New Hampshire Infantry. Source: Child, William. *A History of the Fifth Regiment New Hampshire Volunteers, in the American Civil War; 1861–1865.* Bristol, NH: R.W. Musgrove, Printer; 1893.

find the brother in distress. He was found, placed on a blanket, and at great risk drawn out of range of the rebel rifles, and then carried to the Fifth New Hampshire hospital. He proved to be First Lieutenant Edon of the Alabama volunteers, badly wounded in the thigh and breast. A few hours and he would have perished. Lieutenant Edon informed his brethren of another wounded Mason, who, when brought out, proved to be a lieutenant colonel of a Georgia regiment. These two wounded rebel officers received the same attention as the wounded officers of the Fifth, and a warm friendship was established between men who, a few hours before, were in mortal combat. . . . These Confederate Masons, with some ten Union Masons wounded in the battle, were placed in a barn and for several days were cared for by Assistant Surgeon William Child and Chaplain Ransom, both Masons."[12]

Corroborating the story from a different angle, Union First Lieutenant Janvrin W. Graves of Company H, 5th New Hampshire Infantry, recalled his search for Second Lieutenant Charles W. Bean.

> Right over in front of Dunker church was a rebel battery in position. There I found Charlie Bean, I got them to take him back to the ambulance. I was going back where the guns were and that was where General Richardson got his mortal wound. (Editor's Note: Major General Israel Richardson's wound pierced his internal organs. He was evacuated back to the Pry House [**Chapter 4, Stop 1**] and received treatment from Surgeon John Howard Taylor.) I went back; there was not much fighting after that. When I found our field hospital it was pretty near night. There were twenty-eight of us in the basement, and all but four belonged to our regiment; two to a New York regiment; Colonel Nesbit of the Thirtieth Georgia and Lieutenant John Woden [Oden] of the Tenth Alabama.[13]

"After this battle Assistant Surgeon Child was detailed to remain in charge of the wounded of the brigade on the field. The barns and sheds in all this region were occupied as hospitals by the Union army, and many Confederate wounded were retained here, and I believe were as well cared for as the Union men. The barns were filled with flies, and wounds were sure to gather maggots about the dressings and even within the raw surfaces. To avoid this disgusting evil Assistant Surgeon Child personally gathered a few scores of the shelter tents left on the battlefield, brought them to a suitable location, and with them built very comfortable hospital quarters, and into them moved all the wounded of the Fifth, where they remained until able to be sent to Frederick city or were sent to the Antietam hospital, which was finally established upon the western borders of the battlefield. Child was detailed for service in this Antietam field hospital, where he remained until about December 10. He then returned to his regiment for duty while it was engaged in the battle of Fredericksburg. This Antietam hospital was established for the reception of those who were so severely wounded as not to be able to be removed in the ambulance to Frederick city."[14]

Letter Home

The following is a letter from Surgeon William Child to his wife in September 1862:[15]

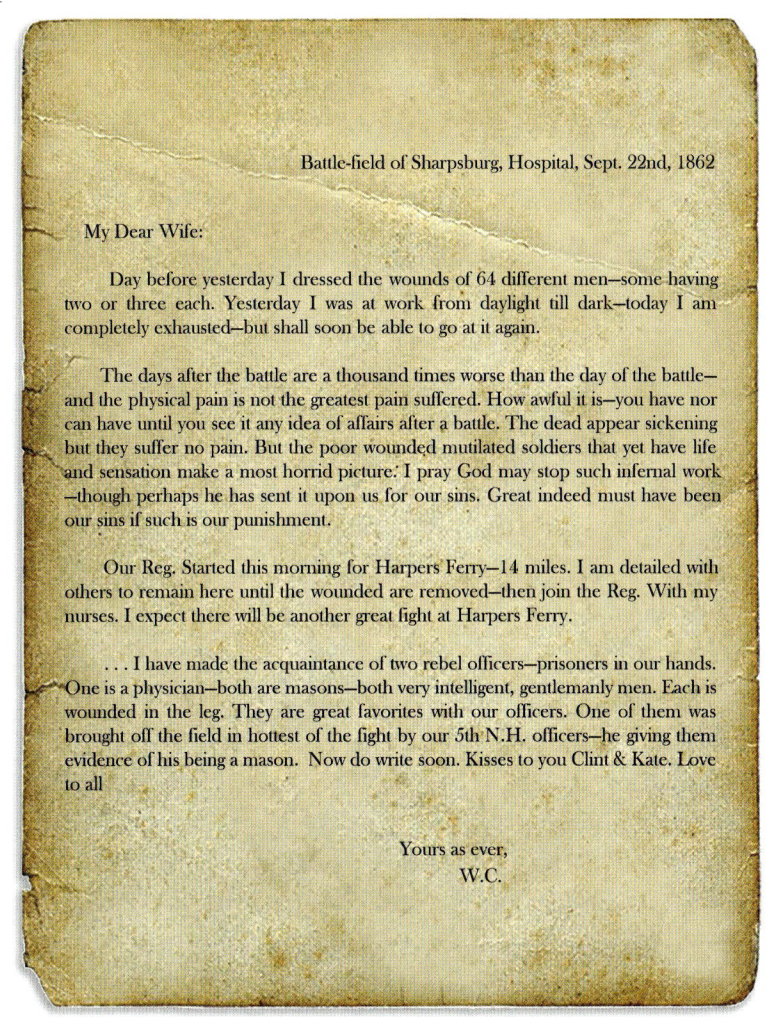

> Battle-field of Sharpsburg, Hospital, Sept. 22nd, 1862
>
> My Dear Wife:
>
> Day before yesterday I dressed the wounds of 64 different men—some having two or three each. Yesterday I was at work from daylight till dark—today I am completely exhausted—but shall soon be able to go at it again.
>
> The days after the battle are a thousand times worse than the day of the battle—and the physical pain is not the greatest pain suffered. How awful it is—you have nor can have until you see it any idea of affairs after a battle. The dead appear sickening but they suffer no pain. But the poor wounded mutilated soldiers that yet have life and sensation make a most horrid picture. I pray God may stop such infernal work—though perhaps he has sent it upon us for our sins. Great indeed must have been our sins if such is our punishment.
>
> Our Reg. Started this morning for Harpers Ferry—14 miles. I am detailed with others to remain here until the wounded are removed—then join the Reg. With my nurses. I expect there will be another great fight at Harpers Ferry.
>
> ... I have made the acquaintance of two rebel officers—prisoners in our hands. One is a physician—both are masons—both very intelligent, gentlemanly men. Each is wounded in the leg. They are great favorites with our officers. One of them was brought off the field in hottest of the fight by our 5th N.H. officers—he giving them evidence of his being a mason. Now do write soon. Kisses to you Clint & Kate. Love to all
>
> Yours as ever,
> W.C.

Editor's Note: Later, Dr Child witnessed the assassination of President Abraham Lincoln in 1865.

From Chapter 1, Stop 2, proceed south toward the Dunker Church (Chapter 1, Stop 3). Along the route, participants will pass the Maryland Monument. It is unique among war monuments because it memorializes units on both sides of the conflict.

STOP 3: DUNKER CHURCH

Figure 1.10 **Dunker Church at the Time of the Battle.** A flag of truce allowed for the caring and evacuation of the wounded near the D.R. Miller Cornfield just north of the Dunker Church in 1862. Source: Library of Congress. Waud, Alfred R. Sharpsburg, MD; 1862.

Figure 1.11 **Dunker Church.** A modern view of the Dunker Church at the Antietam National Battlefield. Today, the scene shows little evidence of the many dead and wounded who were strewn across the grounds in 1862. Courtesy of Scott C. Woodard, US Army Medical Department Center of History and Heritage; November 2017.

During the Battle of Antietam, the Dunker Church (**Chapter 1, Stop 3**) became a beacon of refuge from enemy fire, as well as a reference point in the morning sortie. Facing northeast from the front porch of the church toward the Maryland Monument was Hood's Division (Confederate Lieutenant General John B. Hood), and its troops flowed past like a raging torrent splits a river rock. Hood was pushing to relieve the bloodied division under Confederate Major General Thomas J. Jackson in a measure to hold the D.R. Miller Cornfield. The Federal Army moved in from the east as Sumner's II Corps (Major General Edwin V. Sumner) attacked with Sedgwick's Division (Union Major General John Sedgwick) and pushed the Confederates into the West Woods to the back left. Additional divisions arrived from Harper's Ferry, West Virginia, as the Army of Northern Virginia issued a counterattack into the Federally held cornfield. Sedgwick's Division was repulsed and sent flying into the arms of the Union II Corps rear near the Joseph Poffenberger Farm beyond the initial stop. The guns from French's Division (Union Major General William H. French) could be heard in the distant east as they advanced against Hill's Division that had taken a position on the soon-to-be-bloodied Sunken Road.

Stepping inside the church, one can imagine the scene occupied by Confederate surgeons and hospital stewards. The sulfuric smell wafts from rifle and cannon fire as it mixes with the lead taste of blood now coating the palate. Unhinged doors, litters, and boards span pews and hold the wounded where they may have once prayed.

Diet and Diarrhea

Captain William Parker, a Confederate Virginia Artillery battery commander, whose unit was deployed along the Hagerstown Pike directly across from the Dunker Church, was a physician in civilian life. He recalled that during the march into Maryland his unit was short of rations for three weeks. The soldiers subsisted largely on ears of green corn and orchard fruit, a diet that produced widespread diarrhea. As a result, many of his men suffered from "ulcerated rectums" as they braved Union fire and "served the guns."[16]

Surgery Amid the Guns

Around 6 am on 17 September 1862, Parker's Virginia artillerymen were already hotly engaged near the Dunker Church as the Union advance rolled southward down the Hagerstown Pike. One of Parker's men, Lieutenant Parkinson, fell with a shattered knee. Parker dismounted and hurriedly operated on his prize officer in an effort to save his life. He wanted him back to his guns.[17]

Temporary Medic

Following the repulsion of Sedgwick's division-level attack, the 71st Pennsylvania Infantry launched a counterattack to delay the pursuing Confederate troops. Union Colonel Isaac J. Wistar, 71st Pennsylvania Commander, fell wounded and was left behind when his regiment withdrew during the back-and-forth fight in D.R. Miller's Cornfield. After the war, he wrote:

> . . . I was myself knocked over by a bullet through the left shoulder. Rogers, the left flank sergeant of G Company was instantly at my side, and as the blood was spouting from under the sleeve at the wrist, hastily clapped on a tourniquet constructed of my pocket handkerchief and his bayonet. He offered to remain with me, and was inclined to insist, till I appealed to him to save my sword. Recognizing that obligation, he quickly took it from me, and rushed after the retiring column, and was scarcely gone till the enemy's line marched over me.[18]

A Confederate officer of the 12th Georgia Infantry came across Wistar, who was bleeding heavily and in great pain. The Georgian demanded Wistar's sword and became abusive when informed that Wistar had sent the blade to the rear to prevent its loss. A group of senior Confederate officers observed the action nearby while a courier on Confederate General Jeb Stuart's staff also observed Wistar's distress. Once informed, Stuart immediately dismounted and provided aid. He then sent the irate Georgian on his way.

Wistar relates the following in his memoir:

> The courier then rearranged the tourniquet, which though hitherto but partially effective, had become excessively painful, handed me a drink ... and leaving me in a much more comfortable condition, rode away after his General. It was not till several years after the war that a mutual friend—accidentally hearing the celebrated Confederate guerilla, John Mosby, relate the same circumstance in connection with my name, which he still remembered—brought us together, when I learned for the first time that the friendly courier had been no other than the renowned Mosby, at that time not even a commissioned officer.[19]

Figure 1.12
Colonel Isaac J. Wistar. Union Colonel Isaac J. Wistar, Commander of the 71st Pennsylvania Infantry, was 34 years old at the time of the Battle of Antietam. Wistar, Isaac Jones. *Autobiography of Isaac Jones Wistar (1827-1905) in Two Volumes, Volume II*. Philadelphia, PA: The Wistar Institute of Anatomy and Biology; 1914.

Direct Action on the Battlefield 39

The Wounded Gauntlet

Figure 1.13 **Confederate Line in the Woods.** The Confederate line in the woods near the Dunker Church at Sharpsburg, Maryland. Source: Library of Congress. Waud, Alfred R.; 1862.

Wistar was assisted off the battlefield by two fellow wounded soldiers, but not without dismay.

> During the afternoon, the infantry fighting in our vicinity was mostly suspended, but the thin woods where we lay was severely shelled by the artillery of both sides, tearing to pieces the trees, splintering the rocks and producing terrible results on the helpless wounded of both armies, few of whom in my vicinity survived

it. After dark all regular firing ceased, and some gentle showers gratefully refreshed such as were still alive and able to appreciate them. Two soldiers of the 71st less badly hurt than myself, insisted they could get me off, if I was able to stand, which with their aid I managed to do, but as the ground in our rear was obstructed not only by the multitude of dead and wounded of both armies who here lay thick, but by branches of trees and other results of the heavy artillery-fire so long concentrated on the place, the only available route for three cripples must at first be nearly parallel with the enemy's new infantry line, not fifty yards distant, and with no pickets out. In response to our explanation and request not to fire, they called to us to "go ahead," which precautionary process had to be repeated several times as we passed in front of fresh parts of their line. At last we came to a small farm lane absolutely piled with Confederate dead who had been there mowed down in heaps in repeated but vain efforts to take a Federal battery which had been posted at the head of the lane. It was difficult in our condition to crawl over and through the two fences and these tangled corpses lying between them in every attitude of death, but at last it was by mutual aid accomplished, and we came into a comparatively open field whence the hospital men, fully exposed to the enemy if they chose to fire, were cautiously removing the wounded. These men got us upon stretchers, and by an odd coincidence, struck first upon our own regimental field hospital, set up in a small two-roomed Negro cabin. Amputations and operations were proceeding inside and outside, and the floor was slippery with blood, but place was made for me on the only bed, already occupied by three wounded officers of the 71st, where temporary relief was administered.[20]

Staff ride participants will follow Colonel Isaac J. Wistar back to the division-level hospital in Keedysville, Maryland, later in the monograph (Chapter 4, Stop 4).

Horror at Reel Farm Field Hospital

The original site of a Confederate hospital at the Antietam National Battlefield can be seen from the Dunker Church looking southwest toward the Confederate rear.

David Reel's Farm rests on the east side of Taylors Landing Road, midway between Sharpsburg and the West Woods. Reel, his wife, and six of his seven children fled from the property when fighting became imminent. His eldest daughter, 27-year-old Barbara, remained behind to help care for the wounded at the field hospital established in and around the barn.

As the fighting flared in the West Woods and D.R. Miller's Cornfield, the flow of wounded quickly inundated the site. Perhaps the dust raised by the arrival and departure of ambulances, and the movement of litter bearers and the walking wounded, attracted the distant notice of Union gunners atop Elk Ridge to the east. Artillery fire struck the barn, setting it quickly ablaze. The fire spread so rapidly that not all of the wounded could be moved, and an unknown number were burned alive in the inferno of the barn.[21]

Case Study from *The Medical and Surgical History of the War of the Rebellion (1861-65)*

The following case study documents the care provided to a wounded soldier through his final record of care. The Office of the Surgeon General collected all field medical reports and produced the largest compilation of wartime medical care in its time.

Head Wound that was Headstrong

The 2nd United States Sharpshooters were part of the Union 1st Brigade, 1st Division, I Corps, and fought near the Hagerstown Pike in the contest at the D.R. Miller Cornfield.

CASE: Private C. C. Blake, Co. G, 2d United States Sharpshooters, aged 23 years, was struck, at the battle of Antietam, Maryland, September 17th, 1862, upon the top of his head, by ball and buckshot, the missiles passing laterally over the skull. Temporary symptoms of concussion followed, and after lying down fifteen or twenty minutes, the patient walked to a field hospital, a short distance to the rear. His lower extremities, especially the left, were numb. The same sensation existed in a slight degree in the arms. The wound of scalp was two inches long by one inch wide, and fracture of the skull not suspected. The head was shaved and cold water dressings were applied. At the expiration of forty-eight hours, the man started and walked to Frederick, a distance of twenty miles. At the hospital there, a portion of felt from his hat and some hair were removed from the wound. The patient was then sent to Washington, and thence, on the 24th, he was again transferred and arrived at DeCamp Hospital, David's Island, New York, on the 28th. A fissure of the right parietal bone, near the sagittal suture, was discovered. At the expiration of a week, an incision was made by Acting Assistant Surgeon E. B. Root, and some small portions of the external table were removed; the fissure was found to extend upward of two inches beyond the line of the incision. Five [medicinal] clays, and subsequent portions of both tables were removed, exposing the dura mater to the extent of the size of a ten cent piece. The internal table, which was found depressed about four lines, was elevated. The patient suffered bad from neuralgic pain over his eyebrows, extending through the right temple to the wound. These pains and the numbness of the extremities disappeared after the elevation of the depressed bone. The patient was discharged from the service on November 3d, 1862. The wound had nearly healed, there being a few granulations at its centre. These moved with the pulsations of the brain. No head symptom existed. A communication from the Commissioner of Pensions, dated January 2d, 1868, states that Blake is a pensioner, and that his disability is rated total. The care is reported by Surgeon S. W. Gross, U. S. V.[22]

From the Dunker Church (Chapter 1, Stop 3), staff ride participants can travel along the road heading northeast to the East Woods, where Union and Confederate casualties were collected in the many homes and barns surrounding the battlefield.

References

1. Schell FH. Sketching Under Fire at Antietam. New York, NY, and London, England: *McClure's Magazine*, S. S. McClure Co;1904:421–422.
2. Schell FH. Sketching Under Fire at Antietam. New York, NY, and London, England: *McClure's Magazine*, S. S. McClure Co;1904:425.
3. Schell FH. Sketching Under Fire at Antietam. New York, NY, and London, England: *McClure's Magazine*, S. S. McClure Co;1904:429.
4. Frassanito WA. *Antietam: The Photographic Legacy of America's Bloodiest Day*. New York, NY: Charles Scribner's Sons; 1978:150–153.
5. Greiner JM, Coryell JL, Smither JR. *A Surgeon's Civil War: The Letters and Diary of Daniel M. Holt, M.D.* Kent, OH: The Kent State University Press; 1994:28.
6. Priest JM. *Antietam: The Soldiers' Battle*. Shippensburg, PA: White Mane Publishing Co; 1989:37,69–70,305.
7. Priest JM. *Antietam: The Soldiers' Battle*. Shippensburg, PA: White Mane Publishing Co; 1989:58.
8. Priest JM. *Antietam: The Soldiers' Battle*. Shippensburg, PA: White Mane Publishing Co; 1989:62,69.
9. Frassanito WA. *Antietam: The Photographic Legacy of America's Bloodiest Day*. New York, NY: Charles Scribner's Sons; 1978:194–195.
10. Raab SS. *With the 3rd Wisconsin Badgers: The Living Experience of the Civil War Through the Diaries of Van R. Willard*. Mechanicsburg, PA: Stackpole Books; 1999:91.
11. Raab SS. *With the 3rd Wisconsin Badgers: The Living Experience of the Civil War Through the Diaries of Van R. Willard*. Mechanicsburg, PA: Stackpole Books; 1999:xii–xiii,96–97.
12. Child W. *A History of the Fifth Regiment, New Hampshire Volunteers, in the American Civil War, 1861–1865*. Bristol, NH: R.W. Musgrove, Printer; 1893:335–336.
13. Child W. *A History of the Fifth Regiment, New Hampshire Volunteers, in the American Civil War, 1861–1865*. Bristol, NH: R.W. Musgrove, Printer; 1893:113.
14. Child W. *A History of the Fifth Regiment, New Hampshire Volunteers, in the American Civil War, 1861–1865*. Bristol, NH: R.W. Musgrove, Printer; 1893:129–130.
15. Sawyer MC. *Letters from a Civil War Surgeon: The Letters of Dr. William Child of the Fifth New Hampshire Volunteers*. Solon, ME: Polar Bear & Co; 2001:33–34.
16. Priest JM. *Antietam: The Soldiers' Battle*. Shippensburg, PA: White Mane Publishing Co; 1989:2–3.
17. Priest JM. *Antietam: The Soldiers' Battle*. Shippensburg, PA: White Mane Publishing Co; 1989:35.
18. Wistar IJ. *Autobiography of Isaac Jones Wistar (1827–1905) in Two Volumes, Volume II*. Philadelphia, PA: The Wistar Institute of Anatomy and Biology; 1914:65–66.
19. Wistar IJ. *Autobiography of Isaac Jones Wistar (1827–1905) in Two Volumes, Volume II*. Philadelphia, PA: The Wistar Institute of Anatomy and Biology; 1914:66–67.
20. Wistar IJ. *Autobiography of Isaac Jones Wistar (1827–1905) in Two Volumes, Volume II*. Philadelphia, PA: The Wistar Institute of Anatomy and Biology; 1914:67–68.
21. Frassanito WA. *Antietam: The Photographic Legacy of America's Bloodiest Day*. New York, NY: Charles Scribner's Sons; 1978:267.
22. Otis GA. *The Medical and Surgical History of the War of the Rebellion (1861–65), Part I, Volume II, Surgical History*. Washington, DC: US Government Printing Office; 1870:236.

Readiness Averted: Self-Inflicted Wound

Major (Dr) Jonathan Letterman reported on the operations of the Medical Department of the Army of the Potomac on 1 March 1863 and included the report in his memoirs. He addressed the logistical situation before the Antietam battle and how the best medical plans could easily become thwarted when senior-level decisions are made to adjust logistical support, often deemed myopic in retrospect. Dr Letterman wrote the following:

> Orders were given for the transportation of the Army by water to another part of Virginia, and all the vessels that could be obtained, medical transports as well as others, were pressed into service by the Quartermaster's Department. Rapidity of movement being required, the troops were sent off with scarcely any of their ordinary baggage, the ambulances with their equipments were left behind to be sent after the troops as vessels could be spared for that purpose. A large portion of the medical supplies were also left behind, and, in some cases, every thing but the hospital knapsacks, by orders of colonels of regiments, quartermasters, and others; in some instances without the knowledge of the medical officers, in others notwithstanding their protest. It would appear that many officers consider medical supplies to be the least important in an army; the transportation of their baggage is of much more pressing necessity than the supplies for the wounded; and medical officers have been frequently censured (as they were shortly afterwards) for want of articles required in time of action, when these have been left behind, or thrown upon the roadside by orders they were powerless to resist. From the date of the embarkation of the troops at Fortress Monroe to the period when General McClellan was placed in command of the defences of Washington, I know little, personally, of the Medical Department of the Army of the Potomac, as it was not under my control.[1]

Medical Supply Challenges Before the Battle

Dr Letterman, Medical Director of the Army of the Potomac, said the following in his March 1863 report:

" ... Before leaving Washington, I had ordered a number of hospital wagons from Alexandria, Virginia, which reached me at Rockville, in Maryland, whence they were distributed to the different corps. While at this place, I directed the medical purveyor in Baltimore to put up certain supplies, and have them ready to send to such a point as I should direct. Upon our arrival at Frederick, on September 13th, directions were given for the establishment of hospitals at that place, for the reception of wounded in the anticipated battles, and additional supplies, to a large amount, were ordered to be sent from Baltimore at once. The Confederate troops had been in this city but the day before our arrival, and almost all the medical supplies had been destroyed, or had been taken by them. Just previous to our arrival in Frederick, two hundred ambulances were received from Washington, which I distributed to the corps, as rapidly as the movement of the troops would permit. The failure of the railroad company to forward the supplies caused serious annoyance. The railroad bridge over the Monocacy creek, between Frederick and Baltimore, having been destroyed by the Confederate troops, made it necessary to have all the supplies of the quartermasters and commissary, as well as medical departments, removed at that point. A great deal of confusion and delay was the consequence, which seriously embarrassed the medical department; and not from this cause alone, but from the fact that the cars loaded with supplies for its use were on some occasions switched off and left on the side of the road, to make way for other stores; and some of the supplies, I have been informed, never left Baltimore. . . .[2]

Units operating in the field were unable to manage the large number of supplies issued at the regimental-level. Dr Letterman reported the following on 4 October 1862:

> Experience has shown that the medical supply authorized by the Regulations for a regiment for three months is too cumbrous for active operation, instances being frequent where the whole supply has been left on the roadside. Hereafter, in the Army of the Potomac, the following supplies will be allowed to a brigade for one month for active field service. . . .[3]

The urgent supply deficiencies were assuaged only by relief from nongovernmental agencies, such as the US Sanitary Commission and the US Christian Commission. Were it not for these and other organizations providing logistical support, personified in pioneering nurse Clara Barton, the suffering from both armies would have been frightening. The challenge of supply-chain management was corrected a month after the Battle of Antietam when Dr Letterman established a brigade-level supply table.

Relief by Nongovernmental Agencies

Anna Holstein served as a volunteer nurse in the Army of the Potomac for three years. She described her interactions with critical nongovernmental agencies.

❝ At that early day in the history of the war, we found our noble United States Sanitary Commission here, doing a vast amount of good. From their storeroom were sent, in every direction, supplies to relieve the greatest suffering. And to it, strangers as we were to them, we daily came for articles which we found, in our visits to the hospitals, were most urgently needed, and which our own more limited stores could not furnish. They were as freely given to us for distribution, as they had been in like manner intrusted to them by friends at home.[4]

Chap2RA **Bedside Nurse.** A volunteer nurse provides comfort to a poor soul recovering from the war's cruel cost. Many instances of aid were given solely from the patriotic and humanitarian work of volunteers providing medical materiel and care. Post LM. *Soldiers' Letters from Camp, Battle-field and Prison*. New York, NY: Bunce & Huntington, Publishers; 1865.

The 'Box of Delicacies'

The care package has been a favorite of deployed soldiers for ages. Whether it is a nonprofit organization providing for "any soldier" or the gracious box of love from home, the treasured care package is always a nice surprise.

Letter Home

Union Captain Richard Derby, of the 15th Massachusetts Infantry, wrote home to his family in 1862. He misspelled *Thomsonian* in the letter, which referred to botanical alternatives to traditional medicine at that time. Derby was later mortally wounded at the Battle of Antietam. He was last seen rallying his men in battle:[5]

"FREDERICK CITY, Md., Sept. 13, 1862.

". . . .The "box" has come to hand at last! The lemons were so decayed that you could scarcely tell what they were. The can of raspberry smelt like a bottle of ammonia, and had leaked out a little. It was good luck that the cover did not drop off and spoil every thing. The little crackers were all musty, but the cake was still nice, and the sugar, but probably the tea is infected. You cannot send tea with other articles, unless put in *air-tight* packages. That which you sent before was *clove tea* when I got it. The raisins are nice and very palatable. I have not tried the "corn-starch," but the jelly was nearly eaten at the first opening. The *ginger wine* was *terrible* stuff - regular *Thompsonian* medicine! I had a man attacked with colic just as I opened it, and I administered a dose of it with beneficial effect.

I had got tired of cocoa. It is too heavy for hot weather; but now the mornings are getting cool, I can make good use of two boxes.

One box had some bologna sausages soaked in *Balm of Gilead*, which was in a thin bottle next to them. They were not improved! . . . .

Your affectionate son,

RICHARD DERBY

Chap2RA **Derby Crossing the Potomac.**
Captain Richard Derby is depicted crossing the Potomac River after the Battle of Ball's Bluff in Virginia in October 1861. Hanaford PA. *The Young Captain: A Memorial of Capt. Richard C. Derby*. Boston, MA: Degen, Estes, & Co; 1865.

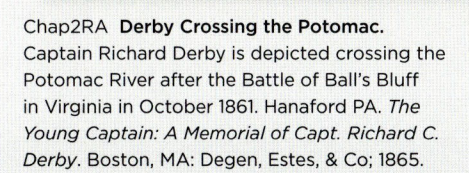

All the Comforts of Home

The following excerpt is from the *Rochester Union and Advertiser Newspaper* of Rochester, New York, on 28 November 1862:

> Smoketown Hospital
>
> The extent of the suffering endured by many of the sick and wounded Union soldiers in the small hospitals of Maryland and Virginia cannot be realized except by a personal inspection of what is going on there.
>
> A young lady of this city who is on a visit to relatives in Maryland, has been amoung the sick and wounded soldiers, and has written a long letter descriptive of what she has seen to a friend here, urging action for the relief of the distressed. She states that she went in company with a lady from Hagerstown—who was devoting much time to the wants of the sick—to a place called Smoketown, about thirteen miles distant, and near the battle field of Antietam. They went in an ambulance, which conveyed many little comforts to the sick. The writer says the lady with whom she went, Mrs. Kennedy, has daily fed hundreds of hungry soldiers, and her house has been filled with the sick and the dying, yet she found time to go thirteen miles distant to minister to those sufferers less fortunate.
>
> Smoketown consists of three small houses and a pig pen. One of the houses is inhabited by a poor family with many children, another is occupied by wounded officers, and contains but two rooms. A school house contains twenty men, and all the rest of the wounded, numbering 600, are in tents, five or six in each. On the arrival of the party they found some; kind ladies administering as best they could to the suffering, and welcomed the coming of more help with joy. The writer says:

"We went into the ladies' tent, while the attendants unloaded the ambulance, and I wish you could have heard the exclamations as the various articals appeared. "Oh, Mrs. K. there is some of your good bread—they got so tired of the hard bread and what the baker brings is so often sour and heavy"! "Onions; we have needed them so much"! "Oh; there are lemons, now we can do so and so," and so on, and I am sure it would have moved you to have seen the eyes of ladies and attendants glisten over the flannel garments, few as they were, and heard their remarks as they took up for instance the grey flannel drawers, (there were only two pairs of them) "won't these be grand for some of those poor fellows who are lying in bed for want of clothes to get up in?" "Won't the men laugh when they see these?" One of the surgeons had just asked imploringly "what they were to do?" The chaplain had taken off some of his own clothing for some one whom he could not refuse."

Many of the wounded were compelled to lie in bed for want of clothing. The writer saw one fellow hobbling about with his crutches, the second time he had been up, arrayed in nothing warmer or more suitable than a pair of light pink calico trousers, and he thought he was fortunate in having these. The men were cheerful, and made the best of what they had.

The description given by the writer of individual cases would be interesting but we have not space for the details. She concludes by urging the people of Rochester to send assistance to the sufferers at Smoketown Hospital. We understand that the ladies of St. Luke's parish have already done something in this way. Supplies may be sent to Mrs. Howard Kennedy, Hagerstown, Md.[6]

CHAPTER 2 READ AHEAD REFERENCES

1. Letterman J. *Medical Recollections of the Army of the Potomac.* New York, NY: D. Appleton & Co; 1866:31–32.

2. Woodward JJ. *The Medical and Surgical History of the War of the Rebellion (1861-65), Part I, Volume I, Medical History, Appendix to Part I, Containing Reports of Medical Directors, and Other Documents.* Washington, DC: US Government Printing Office; 1870:96.

3. Letterman J. *Medical Recollections of the Army of the Potomac.* New York, NY: D. Appleton & Co; 1866:52.

4. Holstein AM. *Three Years in Field Hospitals of the Army of the Potomac.* Philadelphia, PA: J. B. Lippincott & Co; 1867:12

5. Post LM. *Soldiers' Letters from Camp, Battle-field and Prison.* New York, NY: Bunce & Huntington, Publishers; 1865:158–159.

6. National Park Service, US Department of the Interior. *Antietam National Battlefield Lesson Two: One Vast Hospital.* Accessed April 15, 2021. https://www.nps.gov/

CHAPTER 2
Direct Action and Battlefield Hospital

2/1 STOP 1: CLARA BARTON

Figure 2.1 **Antietam Map.** Solid flags represent hospital locations. The "U" represents wounded Union soldiers and the "C" represents wounded Confederate soldiers. This map shows the Samuel Poffenberger Farm location (circled) as treating both Union and Confederate soldiers and, therefore, corresponds with nurse Clara Barton's recollection. Map courtesy of the US War Department. New York, NY: Atlas Publishing Co; 1892.

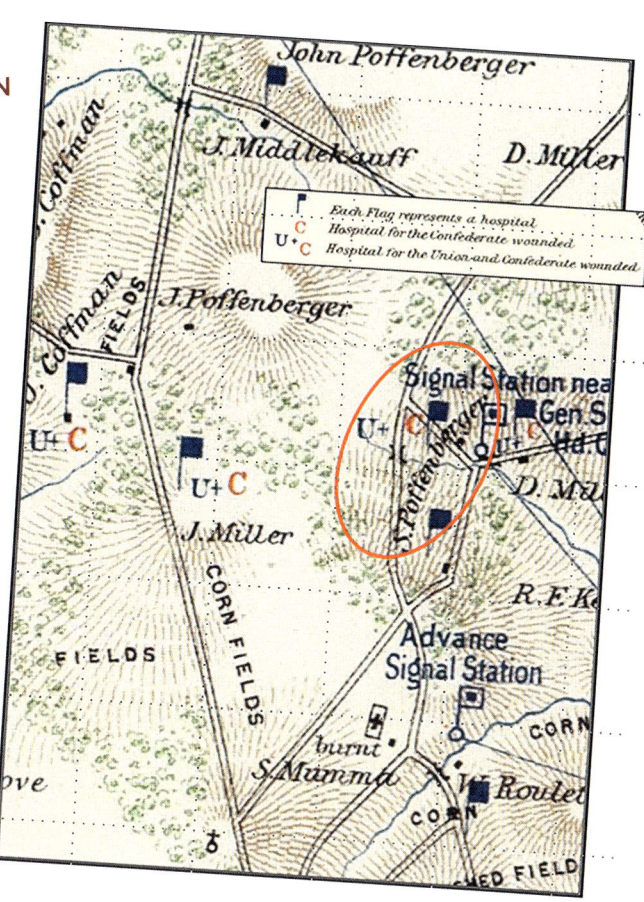

Direct Action and Battlefield Hospital 53

Figure 2.2 [Top] **House and Barn** and Figure 2.3 [Bottom] **House.** The Samuel Poffenberger House and the barn behind it were witnesses to the battlefield in September of 1862. It was here that nurse Clara Barton brought desperately needed medical supplies and treated the wounded. Photographs courtesy of Scott C. Woodard, US Army Medical Department Center of History and Heritage; November 2017.

54 Combat Readiness Through Medicine at the Battle of Antietam

Figure 2.4 **House Drawing.** This photo is a rendering of the original western side of the Samuel Poffenberger House as described by nurse Clara Barton. Many people believe that the present-day western wing of the house seen today was added in 1890 or later. Source: National Museum for Civil War Medicine in Frederick, MD.

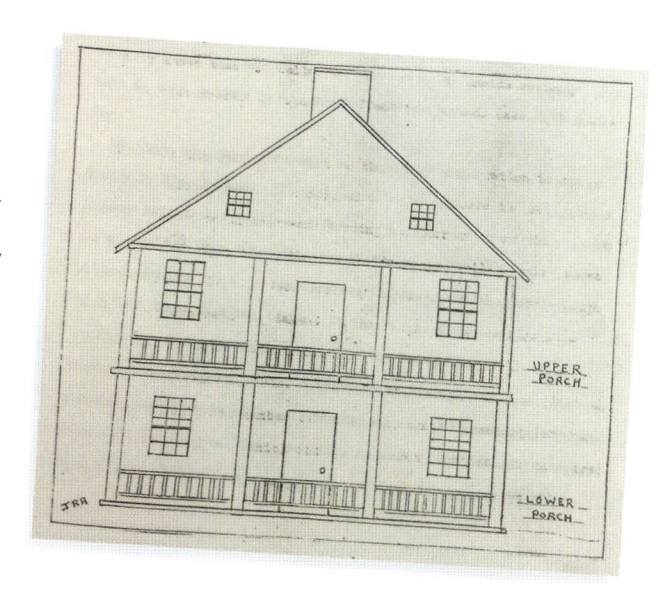

A Small Stone House and Larger-Than-Life Lady

The Samuel Poffenberger House and Barn were in the crossroads of Federal soldiers who were evacuating, mostly from Hooker's I Corps (Union Major General Joseph Hooker). Wounded Confederate soldiers were evacuating from the westward battlefield only a few hundred yards away through the woods. From Mansfield Monument Road (**Chapter 2, Stop 1**) looking north across the farmland, staff ride participants face the stone house described by nurse Clara Barton where she initially arrived on the battlefield (Figure 2.2). Barton previously worked in the patent office in Washington, DC. As a single woman working full-time in the Federal city, her outgoing character and determination pushed the limits of societal expectations. A one-woman fundraising machine, she organized the collection of desperately needed medical supplies for the Union Army and arrived around 10 am with her wagon of medicinal wares during the heat of the morning battle. She greeted her friend, Dr James L. Dunn, when she arrived from the north behind the house.

Dr Dunn explained, "We have nothing but our instruments and the little chloroform we brought in our pockets. Have torn up the last sheets we could find in the house—have not a bandage, rag, lint or string—and all these shell-wounded men bleeding to death."[1]

Figure 2.5 **Clara Barton.**
Wartime photograph of nurse Clara Barton.
Source: Library of Congress. Claflin, Charles R. B.; 1865.

Peering along the western side of the home, Barton wrote, "Upon the porch stood four tables, with an etherized patient upon each. A Surgeon standing over him with his box of instruments and a bunch of green corn leaves beside him."[1]

On 22 September 1862, Missionary Cornelius M. Welles, of the Free Mission Society in Washington, DC, wrote of the battlefield experience he shared with Barton.

> . . . as there was a possibility of a battle, we would prepare a large army wagon load of hospital stores exactly suited to such an occasion. The next day we had our load, as much as four horses could draw. . . . The next day, Tuesday, we were within a mile of our battle lines. Wednesday morning at daybreak, the heavy thundering of artillery on both sides commenced, and the work of destruction and death fairly began. We advanced to the right wing and established our hospital near enough for far-reaching shells to burst over our heads occasionally, but we must be where most needed. Twelve soldiers who had not found their regiments volunteered to assist us. Nearly every little house, barn or stable was already full from previous skirmishing. We had a corn-field by the side of a hay-stack. The hay we used for beds. Soon we were entirely surrounded by those whose wounds were of the most ghastly and dangerous character, legs and arms off, and all manner of gaping wounds from shell and minnie [sic] balls. All such could be removed but a short distance were brought to us.[1]

Never Sewed the Sleeve

A little after 12 pm, Barton described the medical team's position and proximity to the fighting (just to the left and rear) in the East Woods.

> The smoke became so dense as to obscure our sight and the hot sulphurous breath of battle dried our tongues and parched our lips to bleeding. We were in a slight hollow and all shell which did not break among our guns in front came directly among or over us—bursting above our head or burying themselves in the hills beyond. A man lying upon the ground asked for a drink. I stooped to give it and having raised him with my right hand was holding the cup to his lips with my left when I felt a sudden twich [sic] of the loose sleeve of my dress—the poor fellow sprang from my hands and fell back quivering in the agonies of death—a ball had passed between my body and the right arm which supported him—cutting through the sleeve and passing through his chest from shoulder to shoulder. There was no more to be done for him and I left him to his rest—I have never mended that hole in my sleeve. I wonder if a soldier even does mend a bullet hole in his coat?[1]

In the Face of Danger

Clara Barton wrote to the Ladies of the Soldiers' Friend Society of Hightstown, New Jersey, on 14 February 1863, and remarked,

> How our noble surgeons, humane and brave, toiled upon the tables, and we upon the ground, to stay the crimson tide where life was ebbing out, and cheer, and comfort, and restore, and nourish the poor fainting sufferers, until substantial aid could reach them, or receive the last dying message, and soothe as best we might the sudden passage through the dark valley and the shadow of death.[1]

In describing her admiration for the endurance she witnessed among the wounded, Barton relayed one story of a soldier with a Minie ball (a type of muzzle-loading bullet) embedded into the left side of his face.

> It is terribly painful he said won't you take it out?" I said "I would go to the tables for a surgeon" No! No! he said catching my dress—they cannot come to me I must wait my turn for this is a little wound. You can get the ball, there is a knife in my pocket—please take the ball out for me. This was a new call—I had never severed the nerves and fibers of human flesh—and I said I could not hurt him so much—he looked up with as nearly a smile as such a mangled face could assume saying—"You cannot hurt me dear lady—I can endure any pain that your hands can create—please do it—twill relieve me so much" I could not withstand his entreaty and opening the best blade of my pocket knife—prepared for the operation. Just as his head lay a stalwart orderly sergeant from Illinois—with a face beaming with intelligence and kindness and who had a bullet directly through the fleshy part of both thighs—he had been watching the scene with great interest and when he saw me commence to raise the poor fellow's head and no one to support it, with a desperate effort he succeeded in raising himself to a sitting posture. Exclaiming as he did so "I will help do that." And shoving himself along upon the ground he took the wounded head in his hands and held it while I extracted the ball and washed and bandaged the face. I do not think a surgeon would have pronounced it a scientific operation, but that it was successful.[1]

Serendipitous Soup and Stars

Around 2 pm, volunteers came to inform Clara Barton that the last bit of bread had been cut and the last hardtack cracker had been pounded. Hardtack, a simple type of biscuit or cracker, was crumbled with a heavy object to make filling field-ready meals. Besides the hardtack, all that was left were three shipping crates of wine packed in sawdust. Unlike the previous nine crates, the last boxes (to their great surprise) were filled with "American Indian meal" (cornmeal). Barton immediately dispatched a message for kettles and began to boil clean water for the miracle gruel.[1]

That evening in the house, she came across a physician lamenting for the over 1,000 soldiers lying wounded around the Samuel Poffenberger Farm. Having no way to treat them in the dark, he estimated that half of the boys would not see the morning light after succumbing to the darkness around them. Barton had just come from instructing personnel to light the barn with the candles she had delivered and took the haggard doctor "by the arm, and leading him to the door, pointed in the direction of the barn, where the lanterns glistened like stars among the waving corn."[1]

Dr Dunn's Reflections

Dr James L. Dunn wrote about working with Clara Barton earlier in the war. His receptive embrace of her assistance on the battlefield had been solidified by her commitment and abilities. He wrote the following on 25 October 1862:

> I was in a hospital in the afternoon, for it was then only that the wounded began to come in. We had expended every bandage, torn up every sheet in the house, and everything we could find, when who should drive up but our old friend, Miss Barton, with a team loaded down with dressings of every kind, and everything we could ask for! She distributed her articles to the different hospitals, worked all night making soup, and all the next day and night; and when I left, four days after the battle, I left her there ministering to the wounded and the dying. When I returned to the field hospital last week, she was still at work, supplying with delicacies of every kind, and administering to their wants, all of which she does out of her own private fortune. Now, what do you think of Miss Barton? In my feeble estimation, Gen. McClellan, with all his laurels, sings into insignificance, beside the true heroine of the age—"the angel of the Battle-field!"[1]

Editor's Note: Barton would later go on to found the American Red Cross in 1881.

In the days after the battle, Barton stayed to help care for the wounded around the Samuel Poffenberger Barn. One soldier she encountered had refused treatment from the male surgeons present, but finally consented to treatment by Barton after being presented by Surgeon Frank Harwood. The young soldier was Mary Galloway, a native of Frederick, Maryland. She came to Antietam in search of her beau, Union Lieutenant Harry Barnard of the 3rd Wisconsin Infantry. The two met while his regiment was stationed in Frederick in October 1861. When the Union Army passed through Frederick on its way to South Mountain, Maryland, she put on a Union uniform and claimed to be the regimental hospital steward with the ambulance train. She spent the day of the battle searching for his regiment, until she was shot in the neck and shoulder. Barton and Dr Harwood removed the bullet, but could not find any information on her precious Lieutenant Harry Barnard.[2]

Staff ride participants will encounter Lieutenant Barnard later at one of the general hospitals in Frederick.

Advancing south from Samuel Poffenberger's Home (Chapter 2, Stop 1), on the way to the Sunken Road (Chapter 2, Stop 2), participants will pass by the abandoned Mumma Farm on the front left from the road. At this time in the battle, the homestead—set fire by Confederates—was under a billowing black chimney of ash and soot that denied Union sharpshooters critical protection.

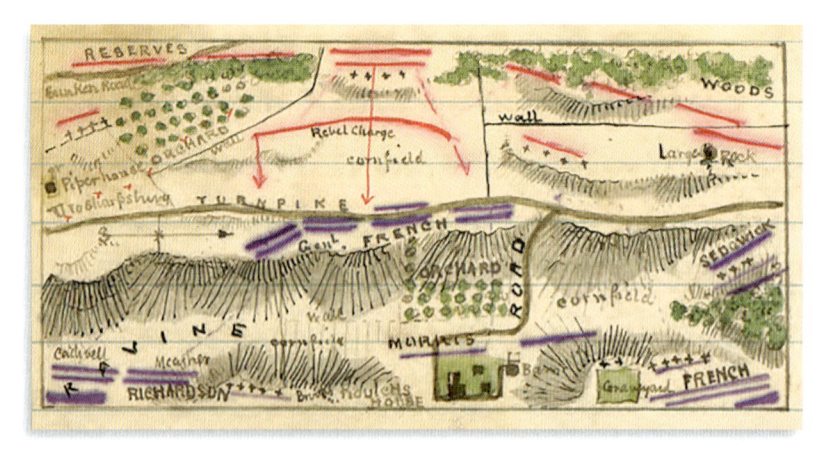

Figure 2.6 **Movement of Federal Troops.**
Cartographer Robert Sneden rendered this visual account of the movement of Federal troops to counter the Rebel push from the Sunken Road at Sharpsburg, Maryland. Source: Library of Congress. Sneden, Robert Knox; 1862.

2/2 STOP 2: SUNKEN ROAD (BLOODY LANE)

Figure 2.7 **Confederate Dead.** The Confederate dead remain in their positions along the Sunken Road at Sharpsburg, Maryland. Source: Library of Congress. Gardner, Alexander; 1862.

Figure 2.8 **Sunken Road Today.** Approximate position of Alexander Gardner's photograph depicting the right wing of the Confederate Army along the Sunken Road at the Antietam National Battlefield, as shown in Figure 2.7. Courtesy of Scott C. Woodard, US Army Medical Department Center of History and Heritage; November 2017.

Direct Action and Battlefield Hospital **61**

Figure 2.9 **300-Meter View.** The Confederate point of view in front of the Sunken Road at the Antietam National Battlefield looking north toward Federal positions. George Wunderlich, Director of the US Army Medical Department Museum (standing 6 feet, 7 inches), is approximately 300 meters in front of the Bloody Lane. The rifled musket was effective within 300 yards (274 meters). Federal soldiers who crested the hill were in the effective range of Confederate infantry rifles. Courtesy of Scott C. Woodard, US Army Medical Department Center of History and Heritage; November 2017.

Standing at the junction of two depressions (**Chapter 2, Stop 2**), the Sunken Road extends left and right when facing north. The second phase of the Battle of Antietam takes place at the center portion of the battlefield. The fighting begins to take shape near the Antietam Creek crossing by elements of Union Major General Edwin V. Sumner's II Army Corps. From this point of the battlefield, Union Major General John Sedgwick's Division begins to push the Confederates back into the West Woods. Union Brigadier General William H. French's Division, while moving toward the fight in the West Woods, was assigned to destroy the gray line at the old Sunken Road. Confederate Major General D. H. Hill and his division were tasked to hold the line to prevent further advances of Federal troops. Urgent reinforcements later arrived from Harper's Ferry, West Virginia, across the Potomac River.

Figure 2.10 **100-Meter View.** The Confederate point of view as the Irish Brigade closes in to take the right flank of the defensive position. George Wunderlich, Director of the US Army Medical Department Museum (standing at 6 feet, 7 inches), is approximately 100 meters in front of the Bloody Lane. If a Federal soldier was able to get this close, the soldier had survived effective fire for the last 200 meters of his advance. Courtesy of Scott C. Woodard, US Army Medical Department Center of History and Heritage; November 2017.

In studying the Irish Brigade's actions upon the Confederate right flank, it is important to realize the bodily damage that came from the Hibernians armed with "buck and ball"—69-caliber ball and 30-caliber buckshot. The Irish Brigade was an infantry unit that consisted mostly of Irish Americans who served in the Union Army. Despite the brigade's tenacity, the entrenched, hidden Rebels poured accurate volley fire into the Irish ranks. The most effective distance for the Sons of Erin was within a 100-meter front, but it was at a staggering loss.

Colonel Gordon's Loving Nurse

Figure 2.11 **Lee and Hill.** General Robert E. Lee, Army of Northern Virginia Commander, and General Daniel H. Hill, Division Commander, are depicted riding along the Confederate lines during a respite in the Battle of Antietam. Source: Gordon, General John. *Reminiscences of the Civil War (Memorial Edition)*. New York, NY: Charles Scribner's Sons; 1904. Atlanta, GA: Martin & Hoyt Co; 1904.

When informed of the vital importance of holding the line to protect General Robert E. Lee's entire Army of Northern Virginia, Confederate Colonel John B. Gordon wrote, "to comfort General Lee and General Hill, and especially to make, if possible, my men still more resolute of purpose, I called aloud to these officers as they rode away: 'These men are going to stay here, General, till the sun goes down or victory is won.'"[3]

At the bend around 10 am, marking the midpoint of Bloody Lane, Gordon's 6th Alabama Infantry weathered the Union Irish Brigade's assault. Hit in the right calf, Gordon ignored the wound as he limped along the firing line. Minutes later, a bullet tore into his leg, just inches above the first wound. Gordon remained with his unit, remembering the promise he made to General Lee. Near noon, a slug tore through his upper left arm. Fighting to stay erect, he barely noticed when a bullet ripped through the crown of his kepi (a forage cap, based on the French kepi). Minutes later, a fourth bullet struck his left shoulder, but he remained on the line. Then a Minie ball struck him squarely in the face below his left eye, exiting on the right side of his neck in a gaping wound. The impact violently snapped his head, hurling his kepi to the ground.

Gordon fell forward onto his face, which lodged in the upturned kepi. Blood from his head and neck wounds flowed into the headgear. If not for the bullet hole in its crown, he would have drowned in his own blood.[4] A litter team carried him from the Sunken Road to a field hospital, where Mrs. Gordon, who had stubbornly insisted on joining him with the Confederate Army, nursed him through his recovery from his multiple wounds. He wrote,

> ... she sat at my bedside, trying to supply concentrated nourishment to sustain me against the constant drainage. With my jaw immovably set, this was exceedingly difficult and discouraging. My own confidence in ultimate recovery, however, was never shaken until erysipelas [infection of the upper dermis], that deadly foe of the wounded, attacked my left arm. The doctors told Mrs. Gordon to paint my arm above the wound three or four times a day with iodine. She obeyed the doctors by painting it, I think, three or four hundred times a day. Under God's providence, I owe my life to her incessant watchfulness night and day, and to her tender nursing through weary weeks and anxious months."[5]

Death of a Gunner Surgeon

Shortly after the Union breakthrough at the Bloody Lane, Confederate Captain M.B. Miller's Battery of the Washington Artillery was deployed in the cornfield at Piper's Farm, which was squarely in the path of the enemy's advance. Although stricken with heavy casualties, the unit stayed in the action only because several officers of Major General James Longstreet's staff assisted in crewing the guns. Dr William Parran, a 29-year-old surgeon of an artillery battalion and former infantry company commander, joined them. He was killed as the battery's soldiers checked the Union advance long enough for a new defensive line to take shape on Piper's Farm. Dr Parran left behind a wife and infant daughter in Barboursville, Virginia.[6]

Figure 2.12 **Union Vantage Point.** The Union Army's point of view looking south toward the Sunken Road at the Antietam National Battlefield. As the photo depicts, the Union soldiers could not see the protected Rebels on the road behind the other side of the slight rise in terrain. Courtesy of Scott C. Woodard, US Army Medical Department Center of History and Heritage; November 2017.

Staff ride participants can travel into the heart of the Federal line across the front of the Sunken Road approximately 300 meters north of the wooden fence. As they look back toward the Confederate line at the fence, they can now see the battle from the Union force's perspective.

2/3 STOP 3: STRAW-STACK HOSPITAL (exact location unknown)

Cared for the Wounded Amidst a Hail of Bullets

The following is an excerpt from *Deeds of Valor From Records in the Archives of the United States Government. How American Heroes Won the Medal of Honor:*

"On the morning of September 17, 1862, the command to which I belonged arrived, after a forced march, on the battlefield of Antietam," Union Assistant Surgeon Richard Curran writes. "My regiment and brigade were immediately put into action. I was the only medical officer present, and, in the absence of orders how to proceed or where to report, I decided to follow my regiment, a course which brought me at once into the midst of a battle, terrible but brief, as the enemy, after a stubborn resistance, yielded, and fell far to the rear. The loss in killed and wounded sustained by the Third Brigade in this charge, and in the subsequent effort to hold the position, was 313.

"The ground of the battlefield at this point was a shallow valley looking east and west. The elevated land on the south was occupied by the Confederates, while the slight ridge on the north was held by our troops and batteries. From this formation of ground, it was impossible for our wounded to reach the field hospital without being exposed to the fire of the enemy. In a battle men will suffer their wounds to go uncared for and undressed for a long time, if in a measurably secure place, rather than expose their lives to obtain surgical

Figure 2.13 **Dr Richard Curran.** Dr Richard Curran, Assistant Surgeon, Union 33rd New York Infantry. Dr Curran was born in Ireland and was awarded the Medal of Honor for his valor on the battlefield. Source: Beyer W.F., Keydel O.F. *Deeds of Valor From Records in the Archives of the United States Government: How American Heroes Won the Medal of Honor.* Detroit, MI: The Perrien-Keydel Co; 1901:81.

Direct Action and Battlefield Hospital **67**

Figure 2.14 **Straw-Stack Hospitals.** Dr Richard Curran performed many brave surgeries close to the fighting front in improvised straw-stack hospitals. Source: Beyer W.F., Keydel O.F. *Deeds of Valor From Records in the Archives of the United States Government: How American Heroes Won the Medal of Honor.* Detroit, MI: The Perrien-Keydel Co; 1901:82.

attention; and this was the case with our wounded. At this point the injured, Union and Confederate, numbering many hundred, preferred to remain close to the ground, and in shelter of the valley, rather than take the risk of seeking care in the rear. During the severest of the fight, and later on, I was told many times by the officers and men, that if I did not seek a place of safety I would surely be killed. I realized that the danger was great, and the warnings just, for, in the performance of my work I had to be on my feet constantly, with no chance to seek protection. But here were the wounded and suffering of my command, and here I believed was my place of duty, even if it cost my life."

Dr Curran continues, "Close to the lines, and a little to the right, were a number of straw stacks. I visited the place and found that many of the disabled had availed themselves of this protection. Without delay I had the wounded led or carried to the place, and here, with such assistance as I could organize, although exposed to the overhead firing of shot and shell, I worked with all the zeal and strength I could muster, caring for the wounded and dying until far into the night. My only fear then was that my improvised straw-stack hospital would catch fire. But we were spared this misfortune and the harrowing scene which would have followed. That there was good reason for this fear is illustrated by one of very many similar incidents. While dressing a wound on the leg of a soldier I turned away to get something to be used in the dressing. On my return I found the leg had been shot off by a cannon ball.

"Happily," the doctor concludes, "in no other position could I have rendered equally good service, for I am confident that, by my action, many lives were saved."

In the report of the commanding officers of the brigade, Dr Curran is mentioned in one place as follows:

"Assistant-Surgeon Richard Curran, of the Thirty-third New York Volunteers, was in charge of our temporary hospital, which unavoidably was under fire. He attended faithfully to his severe duties, and I beg to mention this officer with particular commendation. His example is most unfortunately but too rare."[7]

Medal of Honor

The Medal of Honor is the highest award for valor in the US military and was presented for the first time in 1863. When awarding the Medal of Honor, the President recognizes a service member in the name of the US Congress only when that individual has "distinguished himself conspicuously by gallantry and intrepidity at the risk of his life above and beyond the call of duty."

[From left to right]
Figure 2.15 **Medal of Honor** and Figure 2.16 **Civil War Campaign Medal.**
The Medal of Honor was presented for the first time in 1863; the Civil War Campaign Medal was not issued until 1905. Wyllie, Robert E. *Orders, Decorations and Insignia, Military and Civil.* New York, NY, and London, England: G. P. Putnam's Sons; 1921:48.

Dr Curran's recognition reads,

> The President of the United States of America, in the name of Congress, takes pleasure in presenting the Medal of Honor to Assistant Surgeon Richard J. Curran, United States Army, for extraordinary heroism on 17 September 1862, while serving with the 33d New York Infantry, in action at Antietam, Maryland. Assistant Surgeon Curran voluntarily exposed himself to great danger by going to the fighting line there succoring the wounded and helpless and conducting them to the field hospital.[8]

The Civil War Campaign Medal is considered the first campaign service medal of the US Armed Forces and was awarded to members who had served in the American Civil War between 1861 and 1865. The medal was first authorized in 1905 for the 40th anniversary of the Civil War's conclusion. The blue and gray ribbon denotes the respective uniform colors of the United States and Confederate troops.[9]

A Lady's Witness

The Ladies' Aid Society of Philadelphia solicited goods and monies for wounded Union soldiers. In a report, the ladies summed up their intent with the following words: "To the men of our land belongs the honor of fighting our country's battles; ours is the duty and privilege of ministering to the comfort and relief of our brave defenders when sick, or wounded."[10]

Mrs. John Harris, Secretary of the Ladies' Aid Society, describes the scenes where she nursed soldiers, mainly from Meagher's Irish Brigade (Brigadier General Thomas Francis Meagher):

> Night was closing in upon us—the rain falling fast; the sharpshooters were threatening all who ventured near our wounded and dying on the battle-ground; a line of battle in view, artillery in motion, litters and ambulances going in all directions; wounded picking their way, now lying down to rest, some before they were out of the range of the enemy's guns, not a few of whom received their severest wounds in these places of imagined safety; add to this, marching and countermarching of troops; bearers of dispatches hurrying to and fro; eager, anxious inquirers after the killed and wounded; and the groans of the poor sufferers under the surgeons' hands—and you may form some faint idea of our position on that eventful evening. Reaching a hospital but a few removes from the cornfield in which the deadliest of the strife was waged, I found the ground literally covered with the dead and wounded—barns, hayricks, outhouses of every description, all full. Here and there a knot of men, with a dim light near, told of amputations; whilst the shrieks and groans of the poor fellows, lying all around, made our hearts almost to stand still. The rain fell upon their upturned faces, but it was not noticed; bodily pain and mental anguish—for many were brought to meet the king of terrors face to face, and would have given worlds to evade his cold touch—rendered them indifferent to their surroundings.

Most of the sufferers were from General Meagher's Irish brigade [Kennedy, Neikirk, and Phillip Pry homes[11]], and were louder in their demonstrations of feeling than are the Germans, or our own native born. We could do little that night but distribute wine and tea, and speak comforting words. . . . We remained at this hospital until the evening of the 19th; we had slept a few hours on the straw upon which our soldiers had lain, and upon which their life-blood had been poured out. We prepared tea, bread and butter, milk punch, and egg-nog; furnished rags, lint, and bandages, as needed, and then came on to French's Division Hospital [Otho Smith's Farm near Keedysville[11]], where were one thousand of our wounded, and a number of Confederates. The first night we slept in our ambulance; no room in the small house, the only dwelling near, could be procured. . . .

Passing over the battle-ground of the 9th [Army Corps], such sights as might cause the general pulse of life to stand-still met our eyes.

Stretched out in every direction, as far as the eye could reach, were the dead and dying. Much the larger proportion must have died instantly—their positions, some with ramrod in hand to load, others with gun in hand as if about to aim, others still having just discharged their murderous load. Some were struck in the act of eating. One poor fellow still held a potato in his grasp. Another clutched a piece of tobacco; others held their canteens as if to drink; one grasped a letter. Two were strangely poised upon a fence, having been killed in the act of leaping it. . . .[12]

First 'Paying' Patient at Bloody Lane

The carnage among the attacking Union troops at the Bloody Lane mounted by the second on 17 September 1862. Union Captain Robert A. Abbott of Company G, 132nd Pennsylvania Infantry, spat out a mouthful of blood and teeth after a bullet tore away the bottom of his jaw. Staggering back, he found Assistant Surgeon George W. Hoover, who was fresh out of medical school and joined the Union Army only two weeks earlier. Dr Hoover pulled Abbott behind the shelter of a hay mound and saved his life with immediate emergency surgery. Without the aid of an anesthetic, Dr Hoover performed an operation that afterward was claimed to only be done once before in surgical history—and it saved Abbott's life.[13] Abbott was one of 142 men wounded in the unit that morning.[14]

His Rear Exposed

As the Union 61st New York Infantry approached the Bloody Lane around 12:30 pm, its soldiers had to scale a fence. While climbing, Private Barney Rogers, of Company A, snagged and tore his trousers completely off, leaving him naked from the waist down. Rogers pressed forward with his comrades, inspiring great hilarity in a grim situation. Arriving on the firing line, the half-clad Hibernian fought gamely until a bullet glanced off a rock beneath Rogers' foot and ricocheted off to mangle his big toe. A comrade "watched the bare-buttocked Private . . . crawl off the field like a wounded dog. The middle-aged Irishman clumsily stumbled away on two hands and one leg. The other, he kept extended at full length behind him."[15]

Senior Officer Alcohol Abuse

Around 1 pm on 17 September, the Irish Brigade of the Army of the Potomac assaulted Bloody Lane. Mounted rather unsteadily on his white horse, Union Brigadier General Thomas Meagher led his men forward at the double-quick. As he neared the left flank of the 132nd Pennsylvania Infantry, Meagher toppled from his saddle in a drunken stupor. Getting to his feet, he staggered and reeled about, "swearing like a crazy man," in the words of a 132nd Pennsylvania officer. While disputed, enlisted eyewitnesses believed he was drunk. The Irish Brigade swept past him, but it lost four regimental commanders and hundreds of men in the firestorm at the Sunken Road. Meagher was subsequently listed as "wounded."[16]

Dr Letterman's Plan at Work

It was in this field, north of the Confederate line at the Sunken Road, that the evacuation and treatment plan developed by Major (Dr) Jonathan Letterman proved its worth. By the afternoon of 17 September 1862, the medical officers were executing this plan that was rehearsed and developed for evacuation and treatment of the wounded. Field hospitals were initially set up in the barns and homes around the battlefield just below the crest where the Confederates were bearing down on the Federals. It became fairly obvious that these hospitals were deadly close to the actual fighting. So, the Medical Department of the Army of the Potomac began to move the larger field hospitals back toward Antietam Creek and beyond. Field dressing stations were established downhill from the fighting along the natural flow of the wounded away from the front. Here, soldiers were quickly treated with bandages and administered tourniquets. They were then evacuated by foot using litters or by hoof using ambulance wagons. From this stage in the battle, patient evacuations were coordinated for movement to fixed facilities across the creek, such as the Philip Pry House (**Chapter 4, Stop 1**) where Dr Letterman was busy organizing the medical department. The surgeons and hospital stewards at the Pry House had previously trained with Dr Letterman before the Maryland Campaign in the quest to be ready for the fight. This medical readiness emphasis was now paying off. Using the Ambulance Corps, patients were quickly and effectively moved from the dangers of the intense fighting in front of the Sunken Road to higher levels of care. This "quick win" in this portion of the battle foreshadowed the success and importance of Army readiness for future combat operations with the Army of the Potomac and beyond.

2/4 STOP 4: ROULETTE BARN

Figure 2.17 **Union Viewpoint.** The Federal troops' view looking northwest from their position against the Confederates, who were entrenched near the Sunken Road out of frame on the left. The flow of the wounded traveled down to this lower elevation during the battle. Courtesy of Scott C. Woodard, US Army Medical Department Center of History and Heritage; November 2017.

The natural drift of wounded soldiers found their way to the Roulette Farm by following the lower elevation away from the Bloody Lane. Participants can look northwest and see the Roulette Farm (**Chapter 2, Stop 4**). The barn is in the forefront and the spring house is on the right. The home beyond the barn on the left is where the family hid inside the shelter of the basement during the battle. The white barn on the far ridge is part of the Mumma Farm. During the battle, this abandoned homestead was a pillar of black smoke that was seen for miles. As wounded soldiers sought shelter from the clash of arms, the Roulette Farm seemed to be in a perfect location for obtaining relief. The gradual descent made for an easy downhill path away from the open battlefield.

Staff ride participants can progress from the open terrain toward the Roulette Farm. Onlookers can imagine how the wounded would naturally flow to the point of lowest resistance and away from the pillar of smoke at the Mumma Farm.

Direct Action and Battlefield Hospital

Balls Like Bees

Figure 2.18 **Roulette Barn Exterior.** The exterior of the Roulette Barn seen from the east coming from Bloody Lane at the Antietam National Battlefield. This was the first structure built that offered some protection for the wounded fleeing death at the Sunken Road. Courtesy of Scott C. Woodard, US Army Medical Department Center of History and Heritage; November 2017.

Figure 2.19 **Roulette Barn Interior.** The interior of the Roulette Barn at the Antietam National Battlefield. Notice how the original frame displays signs of shaping with hand tools. Courtesy of Scott C. Woodard, US Army Medical Department Center of History and Heritage; November 2017.

The Roulette family had numerous beehives on their farm, but the ravages of war did quite a number in destroying the busy creatures' abodes. Nests of angry bees and "nests" of dangerous Minie balls (muzzle-loading bullets) upended this once perfect refuge, as members of the 132nd Pennsylvania Infantry began to suffer from both.[17]

The barn afforded shelter from the intense afternoon sun, but failed to stop many of the lead missiles. Inside this dwelling, surgeons and hospital stewards provided comfort and aid to the mass of battle-damaged bodies strewn across the hardwood planks.

Billy Yank Atop Johnny Reb

Union Private J. O. Smith, of the 13th New Jersey Infantry, bore witness to an incredible event at the Roulette Barn Hospital in a letter he wrote soon after the event.

> A strong, sturdy-looking Reb was coming laboriously on with a Yank of no small proportions perched on his shoulders. Wonderingly I joined the group surrounding and accompanying them at every step, and then I learned why all this especial demonstration; why the Union soldiers cheered and again cheered this Confederate soldier, not because of the fact alone that he had brought into the hospital a sorely wounded Federal soldier, who must have died from hemorrhage had he been left on the field, but from the fact, that was palpable at a glance, that the Confederate too was wounded. He was totally blind; a Yankee bullet had passed directly across and destroyed both eyes, and the light for him had gone out forever. But on he marched, with his brother in misery perched on his sturdy shoulders. He would accept no assistance until his partner announced to him that they had reached their goal—the field hospital. It appears that they lay close together on the field, and after the roar of battle had been succeeded by that painfully intense silence that hangs over a hard-contested battlefield; where the issue is yet in doubt, and where a single rifle shot on the skirmish line falls on your ear like the crack of a thousand

cannon. The groans of the wounded Yank reached the alert ears of his sightless Confederate neighbor, who called to him, asking him the nature and extent of his wounds. On learning the serious nature of them, he said: "Now, Yank, I can't see, or I'd get out of here mighty lively. Some darned Yank has shot away my eyes, but I feel as strong otherwise as ever. If you think you can get on my back and do the seeing, I will do the walking, and we'll sail into some hospital where we can both receive surgical treatment." This programme had been followed and with complete success.

We assisted the Yank to alight from his Rebel war-horse, and you can rest assured that loud and imperative call was made for the surgeons to give not only the Yank, but his noble Confederate partner, immediate and careful attention.

J. O. Smith[18]

Case Studies from *The Medical and Surgical History of the War of the Rebellion (1861–65)*

The following case study documents the care provided to a wounded soldier through his final record of care. The Office of the Surgeon General collected all field medical reports and produced the largest compilation of wartime medical care in its time.

Depression from the Top

CASE: Private Jacob Arnold, Co. E, 64th New York Volunteers, aged 22 years, received, at the battle of Antietam, Maryland, September 17th, 1862, a gunshot fracture of the left parietal bone [part of the cranium], with depression of both tables. Treated fast at his regimental hospital, he was sent, on September 24th, to the general hospital at Frederick. On admission, he had complete paralysis of the right leg and arm, and several convulsions soon after occurred. A crucial incision was made, and depressed bone was elevated and removed by Assistant Surgeon R. F. Wier, U. S. A. The flaps were then replaced, and adhesive strips and cold-water dressings were applied. Erysipelas [superficial infection] of the forehead supervened, but this was successfully treated by the usual remedies. By November 17th the wound had cicatrized [healed scar], and by April, 1863, the paralysis had disappeared. Arnold was discharged from the service May 21, 1863. He is a pensioner, and his disability is rated total.[19]

Case Studies from *The Medical and Surgical History of the War of the Rebellion (1861–65)*

The following case study documents the care provided to a wounded soldier through his final record of care. The Office of the Surgeon General collected all field medical reports and produced the largest compilation of wartime medical care in its time.

Mystery Ball in the Ribs

CASE: Private Morris Ward, Co. H, 63d New York Volunteers, aged 32 years, received two gunshot wounds at Antietam, Maryland, September 17th, 1862. One ball entered the back between the eighth and ninth ribs, left side, halfway between the angle and the junction with the costal cartilage, and lodged; the other entered two inches posterior of anterior superior spinous process of the ilium [largest part of the hip bone] of left side and lodged in the gluteal muscles. He was at once conveyed to the field hospital of the Second Corps, where water dressings were applied. On September 30th, he was transferred to the hospital at Frederick, Maryland. Previous to admission, the patient did not complain of cough or pain in the chest. On October 5th, a ball could be distinctly felt beneath the nipple, but as the patient was quite weak from profuse suppuration from the wound in the gluteal region, the removal of the ball by excision was deferred.

October 8th, burrowing of pus among the gluteal muscles and accumulation of gas. An incision was made an inch above the folds of the nates [buttocks]; but little evacuation of pus. Poultice ordered. On the 12th, another incision was made just below the crest of the ilium and a seton [thread or tool to provide drainage] passed through the wound of entrance. The patient seemed much prostrated, but had no cough or expectoration. Tonics and stimulants administered. On the night of the 13th, he expectorated a small quantity of blood, and, on the next day, complained of pain in the chest where the ball had lodged. On examining the chest, a tumor was found to extend from the left nipple downward and inward for two inches, and of the same dimensions in breadth. On applying the ear to the tumor, a sound was heard resembling the passage of air, with a small quantity of liquid, through a slight opening. Tumor tympanitic [distension from air accumulation] on percussion; respiration but slightly embarrassed, but patient very restless. The tumor continued to increase, and the patient failed rapidly, notwithstanding the free administration of stimulants; death occurred on the morning of October 17th, 1862. Necropsy: Rigor mortis well marked. Body considerably emaciated. On laying open the wound on the posterior portion of the chest, the ninth rib was found fractured at that point. On dissecting up the skin over the tumor, some extravasation [leakage] of blood was found underneath. No opening where the ball could have entered the chest could be found, and it seemed probable that it had passed externally, glancing on the ninth rib, yet no external track was visible.

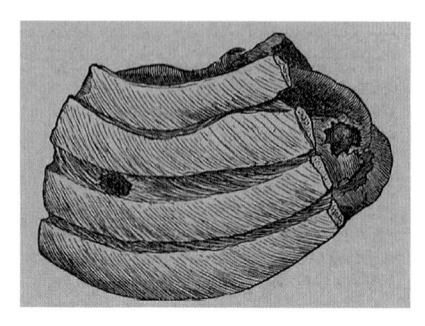

Figure 2.20 **Anterior Ribs.** Segments of the anterior portion of the ribs. They are shown with a condensed portion of the upper lobe of the left lung with a round ball that lay against the diaphragm. Source: US Army Medical Museum.

The pericardium [membrane] over the apex of the heart was adherent to the ribs. Recent pleuritic adhesions on both sides, and left lung adherent to ribs for a large space, where the ball was found underneath. Missile had ulcerated through the intercostal muscle into the lung and was found resting against the diaphragm at the bottom of a large abscess which contained air and pus. Pieces of clothing and bone were also discovered in the diaphragm. The pathological specimen, showing a wet preparation of a portion of the left lung adherent to sections of the third, fourth, fifth, and sixth ribs [represented in the Figure 2.20 woodcut] was contributed, with a history of the case, by Acting Assistant Surgeon Alfred North. [It must be understood that the ball entered posteriorly between the eighth and ninth ribs, on a level with the sixth rib anteriorly. The description of its course is not very clear. If it did not fracture the rib, whence came the bits of bone found imbedded in the diaphragm?].[20]

The Sweetest Meal, Ever

Figure 2.21 **Roulette Spring House.** Both Confederate and Federal soldiers sought desperately needed water at the Roulette Spring House, as viewed here from the western side of the barn. Courtesy of Scott C. Woodard, US Army Medical Department Center of History and Heritage; November 2017.

Figure 2.22 **Still-Flowing Spring.** The steps shown here lead to a spring that flowed during the Battle of Antietam—and it is still flowing. Natural springs were vital to farms and homesteads, and they were lifesaving necessities when tending to the wounded. Courtesy of Scott C. Woodard, US Army Medical Department Center of History and Heritage; November 2017.

Private Benjamin Clarkson, from Company C, 19th Pennsylvania Infantry of Hancock's Brigade (Union Major General Winfield S. Hancock), went back to the spring under the cover of darkness to wash the crusted blood and brains from his uniform and face that were deposited on him from the explosive impact of a cannon ball and the head of a horse. An enterprising friend had absconded some toppled honeycomb and sweet honey from the Roulette grounds. Together with hardtack, it was the sweetest breakfast Clarkson had enjoyed in the Army—even without coffee. The officers had prohibited fires for fear of renewed fighting.[21]

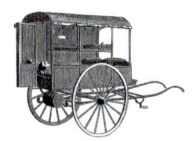

From this location, participants can proceed to the final portion of the fighting at the southern edge of Antietam Creek at the Burnside Bridge (Chapter 3, Stop 1). After crossing the bridge, they can make their way to the eastside of the creek from the parking lot side.

References

1. Atkinson JR. *The Location of the Clara Barton Hospital at Antietam*. Unpublished paper. Frederick, MD: National Museum of Civil War Medicine;1972.
2. Oates SB. *A Woman of Valor: Clara Barton and the Civil War*. New York, NY: The Free Press; 1994:91–93.
3. Gordon JB. *Reminiscences of the Civil War (Memorial Edition)*. New York, NY: Charles Scribner's Sons; 1904:84.
4. Priest JM. *Antietam: The Soldiers' Battle*. Shippensburg, PA: White Mane Publishing Co; 1989:162,169–170.
5. Gordon JB. *Reminiscences of the Civil War (Memorial Edition)*. New York, NY: Charles Scribner's Sons; 1904:91.
6. Frassanito WA. *Antietam: The Photographic Legacy of America's Bloodiest Day*. New York, NY: Charles Scribner's Sons; 1978:211–214.
7. Beyer WF, Keydel OF. *Deeds of Valor: From Records in the Archives of the United States Government. How American Heroes Won the Medal of Honor*. Detroit, MI: The Perrien-Keydel Co; 1901:81–82.
8. Hartke V. *Medal of Honor Recipients 1863–1973: In the name of the Congress of the United States*. Washington, DC: US Government Printing Office; 1973:v,1,69.
9. Wyllie, RE. *Orders, Decorations and Insignia, Military and Civil*. New York, NY, and London, England: G. P. Putnam's Sons; 1921:13,73–74.
10. US Sanitary Commission. *An Appeal to the People of Pennsylvania for the Sick and Wounded Soldiers*. Washington, DC: Philadelphia Associates; 1862:39.
11. US Sanitary Commission. *Report on Field Hospitals indicated on Map of Battlefield of Antietam*. Washington, DC;1862:1.
12. Moore F. *Women of the War: Their Heroism and Self-Sacrifice*. Hartford, CT: S. S. Scranton & Co; 1867:189–194.
13. Hitchcock FL. *War from the Inside*. Philadelphia, PA: J. B. Lippincott & Co; 1904:61,253.
14. Priest JM. *Antietam: The Soldiers' Battle*. Shippensburg, PA: White Mane Publishing Co; 1989:209.
15. Priest JM. *Antietam: The Soldiers' Battle*. Shippensburg, PA: White Mane Publishing Co; 1989:204.
16. Priest JM. *Antietam: The Soldiers' Battle*. Shippensburg, PA: White Mane Publishing Co; 1989:206–207.
17. Frassanito WA. *Antietam: The Photographic Legacy of America's Bloodiest Day*. New York, NY: Charles Scribner's Sons; 1978:200.
18. US Department of the Interior. *Antietam National Battlefield Mumma/Roulette Farm Trail*. Accessed April 15, 2021. https://www.nps.gov/anti/index.htm
19. Otis GA. *The Medical and Surgical History of the War of the Rebellion (1861–65), Part I, Volume II, Surgical History*. Washington, DC: US Government Printing Office; 1870:235.
20. Otis GA. *The Medical and Surgical History of the War of the Rebellion (1861–65), Part I, Volume II, Surgical History,* Washington, DC: US Government Printing Office; 1870:492–493.
21. Priest JM. *Antietam: The Soldiers' Battle*. Shippensburg, PA: White Mane Publishing Co; 1989:294,306.

The Letterman Plan

What we currently call the "Letterman Plan" (named after Major [Dr] Jonathan Letterman) has been the subject of remarkably little critical analysis over the years. Considering the impact of Dr Letterman's far-reaching medical orders of August and October 1862, it is somewhat surprising how misunderstood these orders have been. Informal surveys of officers and enlisted US Army Medical Department personnel have consistently shown that Dr Letterman is rarely seen as an innovator outside of the ambulance system he developed. French and German officers who have been asked similar questions have a far more complete answer. They place primary importance on Dr Letterman's medical organization principles, systematic evacuations, echelons of care, organized treatment near the point of injury, and medical logistics contained in his orders. This is not surprising considering that Dr Letterman's plan was adopted by the Prussian Army in 1865 and was proven worthwhile in the Austro-Prussian War one year later—and then again in the Franco-Prussian War (1870–1871).

It must be understood that Dr Letterman's plan is not entirely unique in either the specific details or in the overall scope. Parts of his plan can be seen in post-Crimean War writings by US Army observers of that war. It can also be seen in the British military's post-Crimea analysis, the US Sanitary Commission's policies and operations, as well as in the works of other officers of the US Civil War such as Major (Dr) Charles Tripler and Major (Dr) Bernard Irwin. What is unique about Dr Letterman's plan was the integration of multiple aspects of medical organization principles, care protocol, logistics, transportation, communications, division of labor, command and control, and a clearer understanding of how the medical mission supports the fighting force.

A careful reading of Dr Letterman's writings shows a clear two-part understanding of the mission. He affirms the overall mission of the Army as paramount. This belief is also reflected in the writings of Dr Tripler who was Dr Letterman's immediate predecessor. In his writings, Dr Letterman begins to define a separate sub-mission for Army Medicine. This sub-mission is to support the overall Army mission while separating certain medical functions from mainstream Army methodologies in order to improve the success of the medical mission. This can be seen in the detailing of specific personnel for stretcher bearer duty; placing ambulances under medical control at the corps, division, and brigade-levels; promoting drill and instruction for ambulance crews separate from their regiments; and in the pooling of surgical personnel into division-level hospitals.

By defining a mission/sub-mission relationship between medical and field commands, Dr Letterman was able to make large scale changes with the support of his superiors and was able to overcome resistance from other Army organizations, such as the Quartermaster Department and even the command staff in Washington, DC. In 1864, much of his plan was passed into law by the US Congress. All of this may seem like a world away from the picturesque landscape of central Maryland.

The Battle of South Mountain on September 14th and Antietam on the 17th were testing grounds for Dr Letterman's fledgling Ambulance Corps and the monumental task of dealing with the highest number of casualties ever taken in the history of our country. These events led Dr Letterman to write two more far-reaching sets of orders on October 4th and October 30th of 1862. In order to develop a full understanding of the development and impact of these orders, it is important to understand the conditions, challenges, and circumstances that shaped them. There is no better way to do this than to study the battle, study the plans, and then follow the campaign on the ground where nearby nearly 23,000 Americans became casualties in a single day.

We may be able to calculate the casualties of this battle, but we will never be able to calculate the number of lives saved by the medical reforms and innovations that it brought about.

> **George C. Wunderlich**
> Director
> US Army Medical Department Museum

Who Has What on the Battlefield?

Major (Dr) Jonathan Letterman was able to create a working structure for his surgeons and their use of medical supplies and equipment. As previously mentioned in the Read Ahead portion of Chapter 2, it was a major shift to no longer push the 90-days-of-supply model and, therefore, overburden troop movements and storage capacity. With the target of 30 days of supply, it was critical that each echelon in the medical service maintain a true readiness posture by exercising strict supply discipline. Providing a smaller quantity of supplies that actually fit storage and transportation capabilities, along with increased resupply efforts, allowed commanders to actually transport needed supplies instead of casting them off in the name of expediency. By clearly defining its procedures, the Army of the Potomac's leadership was able to effectively account for its supplies and their location.

The following correspondence details the medical supplies authorized to a brigade for one month in the Army of the Potomac.[1]

MEDICAL DIRECTOR'S OFFICE,
Army of the Potomac, October 4, 1862.

MEDICAL SUPPLY TABLE FOR THE ARMY OF THE POTOMAC FOR FIELD SERVICE.

Experience has shown that the medical supply authorized by the Regulations for a regiment for three months is too cumbrous for active operation, instances being frequent where the whole supply has been left on the roadside. Hereafter, in the Army of the Potomac, the following supplies will be allowed to a brigade for one month for active field service, viz. :
One hospital wagon, filled.
One medicine chest for a regiment, filled.
One hospital knapsack for each regimental Medical officer, filled.
The supplies in the list marked 'A' to be transported in a four-horse wagon.
The Surgeon in charge of each brigade will require and receipt for all these supplies, including those in the hospital wagon, and will issue to the senior Surgeon of each regiment the medicine chest and knapsacks, taking receipts therefor. The hospital wagon, with its horses, harness, etc., will be receipted for by the ambulance quartermaster.

The Surgeon in charge of the brigade will issue to the Medical officers of the regiments such of these supplies as may be required for their commands, in formally, taking no receipts, demanding no requisition, but accounting for the issues as expended.

The Surgeons in charge of brigades will at once make out requisitions in accordance with these instructions, and transmit them, approved by the Medical Directors of Corps, to the Medical Purveyor of this Army. These supplies being deemed sufficient for one month only, or for an emergency, Medical Directors of Corps will see that they are always on hand, timely requisitions being made for that purpose. Editor's note: The original list is not included in this publication.

These boxes will be kept locked. The Surgeon in charge of the brigade will keep the keys, and, by weekly inspections, ascertain that each ambulance has its full supply. Whenever practicable, one ambulance will follow in the rear of the regiment on the march to transport the medicine chest, knapsacks, and any urgent cases of sickness or wounds. When the ambulance cannot accompany the regiment, one knapsack will be carried by an orderly with the command, and the medicine chest and remaining knapsacks will be placed with the hospital tent and other hospital furniture in the wagon allowed to each regiment for that purpose. The hard bread can always be obtained from the savings of the regimental hospital.

NOTE.-*February 2, 1863*.-In future, in ordinary cases the amount 'on hand' of all articles for which requisition is made must be stated. Requisitions to fill up the brigade supply to the amount ordered to be kept on hand, will be made for such articles as are necessary for this purpose.

    (Signed)   JONA. LETTERMAN,
                                                              Medical Director.

Take Charge of the Wounded Entrusted to Your Care

The significance of detailing the procedures for the establishment of field hospitals may be lost upon today's reader. In the modern military, service members and medical staff take it for granted that a preconceived plan for treatment and hospitalization is provided in a medical concept of support during the combat orders process. These procedures can be found in today's modern operations orders and continue the progress started by Major (Dr) Jonathan Letterman's predecessor (then Major [Dr] Charles Tripler) and peers. Dr Letterman, the Medical Director of the Army of the Potomac, was the trusted and credible staff officer who harnessed command authority. Surgical capabilities in the right place and at the right time with the right leader were put to ink in the aftermath of the Antietam battle. With unit-level detail, Dr Letterman foreshadowed the concept of mission essential tasks. Mindful of the fractious environment swirling around Washington City (nickname for Washington, DC) at that time, Dr Letterman created a decision making process where surgeons collectively made clinical decisions that averted politically motivated second guessing. Membership in this ad hoc committee was based on medical competence, not rank. The group had the authority to decide which surgical procedures were performed. Dr Letterman, an energetic and forward-thinking surgeon, could not be blackballed by jealous career doctors using threats of malpractice as a tool for their own advancement. Most important to the wounded soldier, Dr Letterman threw away the old construct of regiments only serving their own kind. In his view, hospitals were built to treat all wounded soldiers.

The following instructions from Dr Letterman were issued for the organization and staffing of field hospitals.[2]

HEAD-QUARTERS ARMY OF THE POTOMAC,
Medical Director's Office,
October 30, 1862.

[CIRCULAR.]

Sir: In order that the wounded may receive the most prompt and efficient attention during and after an engagement, and that the necessary operations may be performed by the most skilful [sic] and responsible Surgeons at the earliest moment, the following instructions are issued for the guidance of the Medical Staff of this Army, and Medical Directors of Corps will see that they are promptly carried into effect:

Previous to an engagement, there will be established in each Corps a hospital for each Division, the position of which will be selected by the Medical Director of the Corps.

The organization of the hospital will be as follows:

1st. A Surgeon, in charge; one Assistant Surgeon, to provide food and shelter, etc.; one Assistant Surgeon, to keep the records.

2d. Three Medical Officers, to perform operations; three Medical Officers, as assistants to each of these officers.

3d. Additional Medical Officers, Hospital Stewards, and Nurses of the Division.

The Surgeon in charge will have general superintendence, and be responsible to the Surgeon-in-chief of the Division for the proper administration of the hospital. The Surgeon-in-chief of Division will detail one Assistant Surgeon who will report to, and be under the immediate orders of, the Surgeon in charge, whose duties shall be to pitch the hospital tents and provide straw, fuel, water, blankets, etc. ; and when houses are used, put them in proper order for the reception of wounded. This Assistant Surgeon will, when the foregoing shall have been accomplished, at once organize a kitchen, using for this purpose the hospital mess chests and the kettles, tins, etc., in the ambulances. The supplies of beef stock and bread in the ambulances, and of arrow-root, tea, etc., in the hospital wagon, will enable him to prepare quickly a sufficient quantity of palatable and nourishing food. All the cooks, and such of the Hospital Stewards and Nurses as may be necessary, will be placed under his orders for these purposes.

He will detail another Assistant Surgeon, whose duty it shall be to keep a complete record of every case brought to the hospital, giving the name, rank, company, and regiment; the seat and character of injury; the treatment; the operation, if any be performed; and the result; which will be transmitted to the Medical Director of the Corps, and by him sent to this office.

This officer will also see to the proper interment of those who die, and that the grave is marked with a head-board, with the name, rank, company, and regiment legibly inscribed upon it.

He will make out two " Tabular statements of wounded," which the Surgeon-in chief of Division will transmit within thirty-six hours after a battle, one to this office (by a special messenger, if necessary) and the other to the Medical Director of the Corps to which the hospital belongs.

There will be selected from the Division, by the Surgeon-in-chief, under the direction of the Medical Director of the Corps, three Medical Officers, who will be the operating staff of the hospital, upon whom will rest the immediate responsibility of the performance of all important operations. In all doubtful cases, they will consult together, and a majority of them shall decide upon the expediency and character of the operation. These officers will be selected from the Division without regard to rank, but *solely* on account of their known prudence, judgment, and skill. The Surgeon-in-chief of the Division is enjoined to be especially careful in the selection of these officers, choosing only those who have distinguished themselves for surgical skill, sound judgment, and conscientious regard for the highest interests of the wounded.

There will be detailed three Medical Officers to act as assistants to each one of these officers, who will report to him and act entirely under his direction. It is suggested that one of these assistants be selected to administer the anaesthetic. Each operating surgeon will be provided with an excellent table from the hospital wagon, and, with the present organization for field hospitals, it is hoped that the confusion and the delay in performing the necessary operations so often existing after a battle will be avoided, and all operations hereafter be *primary*.

The remaining Medical Officers of the Division, except one to each Regiment, will be ordered to the hospitals to act as dressers and assistants generally. Those who follow the Regiments to the field will establish themselves, each one at a temporary depot, at such a distance or situation in the rear of his Regiment as will insure safety to the wounded, where they will give such aid as is immediately required; and they are here reminded that, whilst no personal consideration should interfere with their duty to the wounded, the grave responsibilities resting upon them render any unnecessary exposure improper.

The Surgeon-in-chief of the Division will exercise general supervision, under the Medical Director of the Corps, over the medical affairs in his division. He will see that the officers are faithful in the performance of their duties in the hospital and upon the field, and that, by the ambulance corps, which has heretofore been so efficient, the wounded are removed from the field carefully and with despatch [*sic*].

Whenever his duties permit, he will give his professional services at the hospital—will order to the hospital as soon as located all the hospital wagons of the brigades, the hospital tents and furniture, and all the hospital stewards and nurses. He will notify the Captain commanding the ambulance corps, or, if this be impracticable, the First Lieutenant commanding the Division ambulances, of the location of the hospital.

No Medical Officer will leave the position to which he shall have been assigned without permission, and any officer so doing will be reported to the Medical Director of the Corps, who will report the facts to this office.

The Medical Directors of Corps will apply to their Commanders on the eve of a battle for the necessary guard and men for fatigue duty. This guard will be particularly careful that no stragglers be allowed about the hospital, using the food and comforts prepared for the wounded.

No wounded will be sent away from any of these hospitals without authority from this office.

Previous to an engagement, a detail will be made by Medical Directors of Corps of a proper number of Medical Officers, who will, should a retreat be found necessary, remain and take care of the wounded. This detail Medical Directors will request the Corps Commanders to announce in orders.

The skilful [sic] attention shown by the Medical Officers of this Army to the wounded upon the battle-fields of South Mountain, Crampton's Gap, and the Antietam, under trying circumstances, gives the assurance that, with this organization, the Medical Staff of the Army of the Potomac can with confidence be relied upon under all emergencies, to take charge of the wounded entrusted to its care.

Very respectfully, your obedient servant,   Jona. Letterman,
Medical Director.

Union Major General George B. McClellan, Commander of the Army of the Potomac, expressed his sincere thanks to the troops under his command in General Orders – No. 160.[3]

GENERAL ORDERS – NO. 160.

HEAD-QUARTERS, ARMY OF THE POTOMAC,
CAMP NEAR SHARPSBURG, MARYLAND, Oct 3, 1862.

The Commanding General extends his congratulations to the army under his command for the victories achieved by their bravery at the passes of the South Mountain and upon Antietam Creek.

The brilliant conduct of Reno's and Hooker's corps, under Burnside, at Turner's Gap, and of Franklin's corps at Crampton's Pass, in which, in the face of an enemy strong in position and resisting with obstinacy, they carried the mountain, and prepared the way for the advance of the army, won for them the admiration of their brethren in arms.

In the memorable battle of Antietam we defeated a numerous and powerful army of the enemy in an action desperately fought and remarkable for its duration and for the destruction of life which attended it. The obstinate bravery of the troops of Hooker, Mansfield, and Sumner; the dashing gallantly of those of Franklin on the right; the steady valor of those of Burnside on the left, and the vigorous support of Porter and Pleasanton, present a brilliant spectacle to our countrymen which will swell their hearts with pride and exultation.

Fourteen guns, thirty-nine colors, fifteen thousand five hundred stand of arms, and nearly six thousand prisoners, taken from the enemy, are evidences of the completeness of our triumph.

A grateful country will thank the noble army for achievements which have rescued the loyal States of the East from the ravages of the invader, and have driven him from their borders.

While rejoicing crowned at the victories which, under God's blessing, have crowned our exertions, let us cherish the memory of our brave comrades who have laid down their lives upon the battle-field, martyrs in their country's cause. Their names will be enshrined in the hearts of the people.

By command of MAJOR-GENERAL M'CLELLAN.
S. WILLIAMS, A. A. G.

Confederate General Robert E. Lee, Commander of the Army of Northern Virginia, expressed his deep thanks to the troops under his command in General Orders – No. 116.[3]

GENERAL ORDERS – NO. 116.

HEAD-QUARTERS, ARMY OF NORTHERN VIRGINIA,
October 2, 1862.

In reviewing the achievements of the army during the present campaign, the Commanding General can not withhold the expression of his admiration of the indomitable courage it has displayed in battle, and its cheerful endurance of privation and hardship on the march.

Since your great victories around Richmond you have defeated the enemy at Cedar Mountain, expelled him from the Rappahannock, and, after a conflict of three days, utterly repulsed him on the plains of Manassas, and forced him to take shelter within the fortifications around his capital.

Without halting for repose, you crossed the Potomac, stormed the heights of Harper's Ferry, made prisoners of more than eleven thousand men, and captured upward of seventy pieces of artillery, all their small-arms and other munitions of war.

While one corps of the army was thus engaged, the other insured its success by arresting at Boonsborough [sic] and combined armies of the enemy, advancing under their favorite General to the relief of their beleaguered comrades.

On the field of Sharpsburg, with less than one-third his numbers, you resisted from daylight until dark the whole army of the enemy, and repulsed every attack along his entire front of more than four miles in extent.

The whole of the following day you stood prepared to resume the conflict on the same ground, *and retired next morning without molestation across the Potomac.*

Two attempts subsequently made by the enemy to follow you across the river have resulted in his complete discomfiture and being driven back with loss.

Achievements such as these demanded much valor and patriotism. History records few examples of greater fortitude and endurance than this army has exhibited; and I am commissioned by the President to thank you in the name of the Confederate States for the undying fame you have won for their arms.

Much as you have done, much more remains to be accomplished. The enemy again threatens us with invasion, and to our tried valor and patriotism the country looks with confidence for deliverance and safety; your past exploits give assurance that this confidence is not misplaced.

R. E. Lee, General Commanding.

CHAPTER 3 READ AHEAD REFERENCES

1. Letterman J. *Medical Recollections of the Army of the Potomac*. New York, NY: D. Appleton & Co; 1866:52–56.
2. Memoir of Jonathan Letterman, M.D., Surgeon United States Army and Medical Director of the Army of the Potomac. *Journal of the Military Service Institution of the United States*. New York, NY, and London, England: G. P. Putnam's Sons; 1883:285–287.
3. *Harper's Weekly*. New York, NY: Harper & Brothers;1862:675.

CHAPTER 3
Last Push

3/1 STOP 1: BURNSIDE BRIDGE

From the east side of the Antietam Creek (**Chapter 3, Stop 1**), the terrain appears very much like it did in September 1862. The large tree, on the right side of the Burnside Bridge of the eastern bank, was a witness to the horrible action that day. Unlike then, it now covers the view of the opposing bank where Georgia infantry aimed their rifles with deadly accuracy. During the Battle of Antietam, the cited bridge was known as the Rohrbach or Lower Bridge. It was later renamed the Burnside Bridge after Union Major General Ambrose E. Burnside.

A Family Dies of the Greatest Killer

Disease and nonbattle injuries continued to take more life than bullets and shells during this conflict. Death's cold grasp did not have to rely upon the weapons of war discharged through black powder and fire. Soldiers often worried of the health of loved ones at home and feared their own pending suffering on the battlefield from their fellow man's destructive nature. On 17 September 1862, around 10 am, as Federal troops began their attempt to storm across the bridge to bring the fight to the Confederates across the Antietam Creek, there were probably many soldiers thinking of their families at home and their relatives in the service. Among the dead at the Burnside Bridge was 18-year-old Private Alvin Flint Jr, Company D, of the Union 11th Connecticut Infantry, who enlisted on 1 October 1861. On 6 December 1861, his mother died of tuberculosis and was followed on 16 January 1862 by his 15-year-old sister, also a victim of tuberculosis. In August 1862, his father, Alvin Flint Sr, and his 13-year-old son, George, enlisted in the 21st Connecticut Infantry. On 10 January 1863, typhoid fever claimed the senior Alvin in camp at Falmouth, Virginia, followed by son, George, on 15 January. In just a little over a year, the entire family had perished—with only the junior Alvin dying in battle.[1]

Figure 3.1 **Burnside Bridge (then).** Artist Edwin Forbes captures the battle scene at the Burnside Bridge. Source: Library of Congress. *The charge across the Burnside Bridge—Antietam.* Forbes, Edwin; September 1862.

Figure 3.2 **Burnside Bridge (now).** A modern view of the Burnside Bridge where New Hampshire and Pennsylvania regiments faced Georgian soldiers across the Antietam Creek. Courtesy of Scott C. Woodard, US Army Medical Department Center of History and Heritage; November 2017.

From Doctor to Private to Hospital Steward

Dr George Bronson volunteered with the 11th Connecticut Infantry and enlisted as a private, despite having a medical degree. After proving his worth, he was entrusted to serve at the rank and position of Hospital Steward. Over time, Dr Bronson proved he was extremely qualified and was able to assist surgeons in a manner beyond his position. His entering voluntarily as a humble enlisted soldier speaks to his character. He wrote to his wife about the aftermath of the battle at Burnside Bridge.[2]

The following is a letter from Dr Bronson to his wife in 1862.[3]

> Sunday Sept. 21, 1862 Sharpsburg, MD
>
> Dear Wife,
>
> Your letters 3 in number reached me 1st evening, and it gave me much pleasure to hear from you. I should have written you before, but did not know for a certainty where to direct. You will doubtless have learned the details of this great battle before this reaches you. The loss of the 11th is dreadful.
>
> I followed in the rear of the Regt. Until it reached the fatal bridge that crosses the creek, this bridge is composed of 3 stone arches and the stream is about the size of that one just west of Berlin. The enemies sharpshooters commenced the action being posted in trees and under cover of a wall on the high ground on the other side of the creek, the order was for the 11th to take and hold the bridge until the division of Genl. Rodman passed.
>
> The action soon became general all along the lines, language would fail me to describe the scene. I was in company with the surgeons and we laid ourselves down between the hills of corn and in a lot west of the bridge being a corn field. I had a bag of bandages and some few other things in hand, we lay low I can assure you and the way the bullets whistled around us is better imagined than described. The shells also bursting over our heads and on the ground around us. The attack was perfectly successful, we fell back to a brick house a mile in the rear and established a hospital.
>
> I took off my coat to dress wounds and met with a great loss. Some villain rifled my pockets of several packages of medicine, my fine tooth comb and what I valued most my needle book containing the little lock of hair you put in. No money would have bought it. It was not the value that I cared for, but the giver. Can you replace it. I should be pleased with your photograph which you spoke of. I think that it will be so that I can get a little box by express soon. I am still in the hospital near the battle ground the Regt. having moved about 3 miles. I will tell you where to send the box soon. You need not put Co. K on my letters in future, but simply Dr G. Bronson 11 Regt. C Burnside division with name of place (Washington) for the present.
>
> Give my love to all our friends.
> Very Truly yours
> *George*

Case Studies from *The Medical and Surgical History of the War of the Rebellion (1861–65)*

The following case study documents the care provided to a wounded soldier through his final record of care. The Office of the Surgeon General collected all field medical reports and produced the largest compilation of wartime medical care in its time.

With the Assistance of Two Comrades

CASE: Private Orrin C. Spencer, Co. F, 11th Connecticut Volunteers, aged 18 years, was wounded at the battle of Antietam, Maryland, September 17th, 1862, by a musket ball which fractured the outer table of the frontal bone at its superior portion and to the left of the median line. He was stunned, but after reaction, endeavored to walk, but was too faint and giddy to go far. With the assistance of two comrades he retired to a field hospital where cold water was applied to the wound. He was transferred to Frederick, and thence to Washington, entering Capitol Hospital on the 22d. On the 24th he was sent to the DeCamp Hospital, David's Island, New York Harbor, where he arrived on the 28th. The wound was discharging freely. At the expiration of a week erysipelatous [local inflammation] action set in, which was, however, readily combatted by a purge and the local application of iodine. On October 26th two pieces of the outer table of the frontal bone were removed. At times he suffered severe pain over his eyebrows which extended over the left side of his head, and occasionally he was so dizzy that he could not walk across the ward. He was discharged from the service on November 12th, 1862. The wound had healed, but dizziness occasionally recurred. On January 3d, 1868, the Commissioner of Pensions stated that Spencer was a pensioner, that his disability was rated at one-third, and the prognosis of its duration doubtful. Surgeon S. W. Gross, U. S. V., reports the early history of the case.[4]

Operating Under Fire

Figure 3.3 **Burnside Bridge—September 1862.** Acting Surgeon Theodore Dimon, of the Union 2nd Maryland Infantry, was able to first view the Burnside Bridge at Sharpsburg, Maryland, from this angle. Source: Library of Congress. Gardner, Alexander; September 1862.

Acting Surgeon Theodore Dimon, of the Union 2nd Maryland Infantry, was with his unit in the fight for the Burnside Bridge along the eastern side of Antietam Creek on 17 September 1862. Looking at the terrain around him from the south side of the bridge, Dr Dimon established a brigade field hospital using the resources at hand. Dr Dimon recalled the following:

> I was taking observations for a place for the wounded and noticed a barn built of round poles and covered with thatch just to the left of me, and a pool of good-looking water just in front. I unhitched my tin cup and stooped to take a drink of the water. Just at this instant, the enemy poured in their volley. It had seemed comparatively still before that, but now it seemed as if all the noises in the world had broken out at once. The batteries were pretty close down on both sides and they all opened. I noticed some splendid practice of our batteries, their shells bursting just at the willows and limbs of them dropping into the stream,

> cut off by their fragments. There were a few loose men standing near me, coming from where I do not know, perhaps skirmishes. I sent them to the barn with directions if it had straw or hay in it, to pile that up against the poles and the front door towards the creek to make a barricade of it against bullets and cover for the wounded, and to take along all the water from the pool their canteens would hold.[5]

Georgian troops were positioned across the creek and fired down into the natural channel created by the bridge. Their deadly accuracy dropped blue uniforms all over the field. Dr Dimon demonstrated his willingness to ignore territorial claims and the randomness of war.

> My barn was crammed very soon and Dr. [Charles T.] Reber of the 48th Pennsylvania, a good surgeon and plucky and cool, came in to help me. A singular thing happened here. Our amputating table consisted of a small door mounted on two barrels. I had just severed the Captain's leg . . . and I had an artery of the stump in a pair of forceps, and Reber was adjusting the ligature on it, when two fragments of one of the numerous shells that were bursting over and coming through our thatched roof came down between our heads and hands and the stump, without touching anything, and plunged into the blood and straw at our feet. (Reber afterwards hunted them out and washed them, and we each kept one as a memento of Burnside's Bridge.) The men I sent up in the beginning found the barn filled with straw in bundles and made a very good barricade against bullets from the front but, of course, all large missiles would come through above. Fortunately, no shell burst in the barn and it did not take fire.[6]

In addition to the expected cases of wounded soldiers from rifles and cannon, a different sort of patient emerged from the melee.

> One man was brought in who was in an uncontrollably restless state, constantly throwing himself about unconsciously onto oth-

ers lying beside him. I had to strip him and wash him, for he was covered with faeces, to find his wound, when, behold, he had no wound discoverable, and no bone broken and yet he looked as he was dying. Stimulants had no effect and he died in half an hour. This was, doubtless, one of those cases that used to be called "windage." He was doubtless struck on the chest or over the liver or abdomen very obliquely by a cannon shot which, though it does not tear the clothes or show marks upon the skin, yet the parts underneath are found in a state of disintegration.[7]

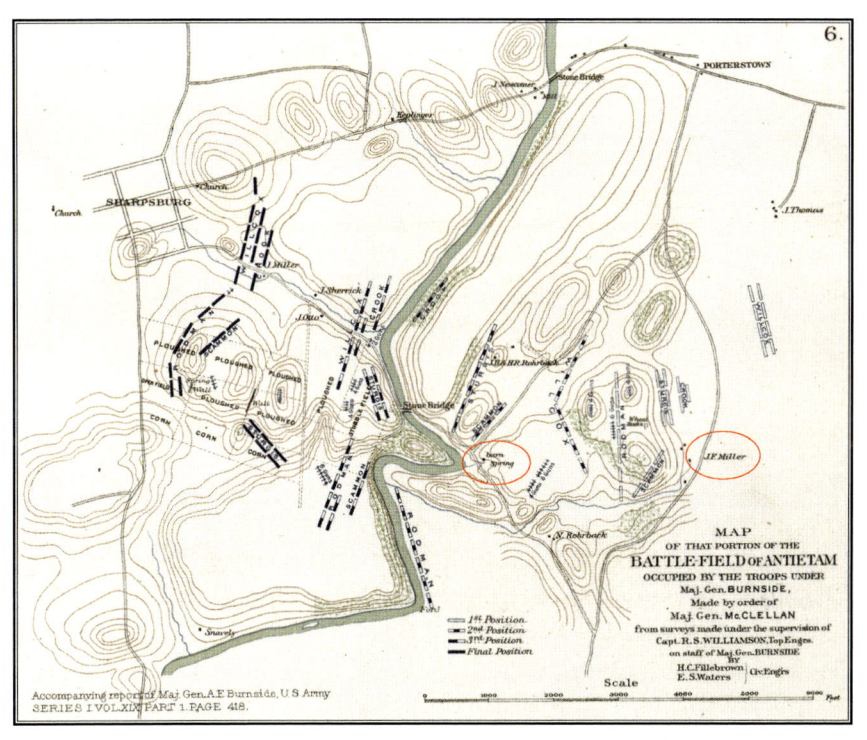

Figure 3.4 **Antietam Map.** Located outside the Antietam National Battlefield, one circle designates the location of the barn and spring used by Acting Surgeon Theodore Dimon of the Union 2nd Maryland Infantry. Source: Library of Congress. US War Department. Davis GB, Cowles CD, Caldwell JA. New York, NY: Atlas Publishing Co; 1892.

Mindful of the tactical considerations of his present situation, Dr Dimon moved the wounded further to the rear and out of range of the artillery.

> I sent one of the men at the pool with a note back to the Assistant Surgeon for my instruments and dressings, first telling him he had better wait where he was with the men and stretchers till it was settled whether we had carried the bridge. The man did his duty faithfully and returned through the cornfield in a few minutes with what I sent for. The action, so far as we were concerned, began about 11 A.M., for I was looking at my watch just before I went down and it was then just 11. About two, as we had finished dressing all in the barn and provided for them as well as we could, I went out to look around. The firing had held up in our vicinity and gone over the other side of the river.

In that 3-hour period, the surgeons and hospital personnel saw about 150 Union soldiers killed and wounded of the 400 souls who were present at the morning report. The hospital was actually a barn built of round poles and covered with thatch.[8]

Orders were given to move all the wounded from the makeshift hospital away from the range of Confederate batteries across the creek. Hospital personnel, using ambulances and stretchers, were able to move and evacuate all of the patients by 3 pm. Surgeon Dimon traveled back along the evacuation route and began to check on the welfare of his patients. At a spring, he came upon a home with an attached barn filled with patients but without a surgeon or supplies. Stragglers—calling themselves "sick"—loitered around the barn and ambulance and made themselves very comfortable in the barn.[9]

Parade of the Maimed

As the fighting for the Burnside Bridge raged on from about 10:30 am to noon, the rookie Union troops of the untried 35th Massachusetts Infantry were held in reserve in the cover of a cornfield not far from the brigade field hospital. Their position lay athwart the path taken rearward by many of the wounded as they sought medical aid.

The beleaguered wounded methodically trotted and felt their way past the untested newbies. It was a parade of the maimed—bloody rags, stumps where legs had been, ripped flesh that once held sleeves. The New England troops were "front row" to a cornfield that ate soldiers and spit out their dying bodies. The tunnel of death from which they emerged would soon swallow them up whole, and some with pity, would be spit out.[10]

Figure 3.5 **Union Soldier at the Rock Wall.** An unknown Federal soldier stands along a rock wall near the Burnside Bridge where many Union soldiers are buried. Source: Library of Congress. Sharpsburg, Maryland. *Graves of Federal soldiers at Burnside Bridge.* Gardner, Alexander; 1821–1882.

Figure 3.6 **Burnside Bridge—November 2017.** George Wunderlich, Director of the US Army Medical Department Museum, shares the scene with the wood and rock that bore witness to the battle. Federal graves were moved continuously up until September 1867 when the cemetery was officially dedicated as the Antietam National Cemetery. The cemetery holds mostly Federal troops; Confederate troops were reinterred and buried at other locations. Courtesy of Scott C. Woodard, US Army Medical Department Center of History and Heritage; November 2017.

Will You Give Us Our Whiskey?

Upon dissuading any notion of fording the Antietam Creek at the bend, Union Colonel Edward Ferrero, 2nd Brigade Commander of the IX Corps, called to his men, "It is General Burnside's special request that the two 51sts (51st New York and 51st Pennsylvania infantries) take the bridge. Will you do it?"

War does many things to soldiers and the fear of death along a humid creek is no different. A known Pennsylvanian teetotaler, Corporal Lewis Patterson, from Company I, broke the silence and asked, "Will you give us our whiskey, Colonel, if we take it?"

Figure 3.7 **Bridge Battle Drawing.** The Burnside Bridge was paid for with a price. Source: Library of Congress. *Battle of Antietam, Maryland—Burnside's division carrying the bridge over the Antietam Creek, and storming the Rebel position, after a desperate conflict of four hours, Wednesday, September 17 / from a sketch by our special artist, Mr. Edwin Forbes.* Forbes, Edwin; 1862.

The affirmative promise to provide the two regiments what they wanted, and the follow-up on whether they accepted the challenge, received a resounding "Yes" from the troops. The wounded and dying lay as testimony to Union Major General Ambrose E. Burnside's personal insistence on a frontal assault, even though he had no personal knowledge of the terrain. The few Southerners in hardened positions had easily thwarted all previous disjointed attacks.

All around the 2nd Brigade was evidence of the earlier attempts to cross the bridge. For about an hour until 1 pm, the struggle continued as the Georgians' ammunition and men were depleted. The Federals began to cross piecemeal above and below the bridge because it was impassable with stacked bodies. The brigade commander had to provide two barrels of whiskey.[11]

An Empty Trouser Leg Emphasizes His Story

Figure 3.8 **Battle Sketch.** The final assault across the Burnside Bridge. Note the smoldering Mumma Farm in the distance. Source: Library of Congress. *The charge across the Burnside Bridge—Antietam / E. Forbes.* Forbes, Edwin; September 1862.

In a newspaper article published in *The Weekly Banner* in Athens, Georgia, in 1892, Private George Washington Lafayette Ard, Company K, 2nd Georgia Infantry, explains how he was wounded near the lower bridge (Burnside Bridge).

> I belonged to the Second Georgia, Toombs' brigade [Confederate Brigadier General Robert A. Toombs]. I was on the extreme right of a few of us who were attempting to prevent Burnside [Union Major General Ambrose E. Burnside] from crossing the lower stone bridge. The fight was on; a ball passed through my thigh, and, while lying on the ground wounded, another ball passed through my right elbow joint. Our forces retreated and the Federals rushed across the creek. Wounded, bleeding, suffering as I was, it was a rare sight to see thousands of well-fed, well-clad soldiers occupying the ground just abandoned by the few ragged, hungry Confederates. The contrast struck me. A regiment of Federals halted near where I was lying. The officer made his men

a short speech, which was cheered. Amidst this, I beckoned to an officer near me and requested that he would drag me on the other side of a tree hard by. He at once stepped back to the line and brought four men, who gently picked me up and placed me behind the tree, hastily spreading a blanket for me to lie upon. I requested to know whom to thank for the kindness. The reply was, "We belong to the Ninth New York Regiment, Hawkins' Zouaves.' These four men hurried back to their places, and the command came from head of column, 'forward, march,' and Burnside's corps passed by.

Very soon an army surgeon came near me. I called to him. Fortunately, I was a Mason, for he was one. He said his name was [George] Humphries, surgeon of the Ninth New York Regiment [Hawkins' Zouaves]. Dr. [Truman H.] Squires, his assistant [89th New York Infantry], was with him. I asked the surgeon if he could give me any temporary aid, remarking that he had as many of his own across the creek as he could attend to. His reply was that he was under as many obligations to me as to any man. He said he had been a surgeon in the Crimean War. He examined my wounds. He administered chloroform, and when I became conscious my leg was off and my arm bandaged.

In that fix I lay behind the tree. The shot and shell from the Confederate batteries were felling treetops and tearing up the ground all around me. Just before night, the firing ceased and the assistant surgeon, Dr. Squires, returned to me and stitched up the flaps of the amputated limb. There I spent the long night. My sufferings, mental and physical, were agonizing. The weather was hot. Loss of blood created thirst. Nearby, I could hear the rippling Antietam mocking me as I called aloud for water which came not.

As a last resort for water I used the grand hailing signal of distress. Some Yankee soldier heard my cry and filled my canteen with water from the creek.

The next morning about sunrise, an ambulance came for me, sent by Dr. Humphries, and took me some two miles to a farmhouse, where Dr. Humphries most tenderly cared for me. He brought a young man who he called Mac and said, 'Mac, I commit this young Georgian, and others to you.'

I found this 'Mac' to be Paul J McLocklin of the Ninth New York Regiment.

In some two weeks, we were removed to a field hospital. My friend Mac continued to wait on me as long as I remained, until the 24th of January, 1863. A nobler man than Paul J McLocklin never lived. While in the hospital, I became acquainted with several members of the Ninth Regiment, and was under the charge of Dr. Humphries until he left for the front, and Dr. Squires was put in charge. In time, I was moved to Frederick City, and I missed the men of the Zouaves.

On the 16th day of May, 1863, I was taken from Frederick City to Baltimore, thence to Fort Norfolk, thence to Fortress Monroe. Here I was transferred to a large steamer, the "Willow Leaf," and the guards on board were Ninth New York men. I was rejoiced. One-legged and maimed I was troubled to know when I reached City Point how I should climb the hill to reach the train that bore the exchanges to "Dixie," but the Ninth New York Zouaves saw me through on board the train.

After the war, Mac and I kept up a correspondence for many years. His letters ceased to come. I wrote again "to be returned to Lumpkin Georgia, if not called for in ten days." The post master at West Winsted, Conn. wrote back, "Your friend Mac died a few months ago." In the meantime, we had exchanged photographs and for years his picture has been hanging upon the wall in my bedroom. If I had money, I would go to Athens, for I want to see the men of the Ninth New York Regiment as I would my own Confederates."[12]

There's Always One Who Doesn't Get the Word

Unlike the rest of the Army of the Potomac, Burnside's IX Corps (Union Major General Ambrose E. Burnside) had not trained and rehearsed the evacuation and ambulance corps drills established by Major (Dr) Jonathan Letterman before the Maryland Campaign of 1862. Medical care during river crossing operations was always a difficult maneuver and competed with the need to carry out sustained firepower. Choices were made on whether to move the wounded from the objective back across the water to medical treatment areas or to push the treatment capabilities across the water with the attacking forces. In one scenario, the lines of evacuation were pushed back against advancing infantry and ammunition. In the other, precious medical assets were in danger of being eliminated. Dr Theodore Dimon, of the Union 2nd Maryland Infantry, demonstrated strong tactical skills when he sought how best to support the action medically. His use of the terrain and knowledge of the battle helped the entire corps who had limited medical planning experience. This situation would be the last time components of the Army of the Potomac were not medically ready for battle.

Case Studies from *The Medical and Surgical History of the War of the Rebellion (1861–65)*

The following case study documents the care provided to a wounded soldier through his final record of care. The Office of the Surgeon General collected all field medical reports and produced the largest compilation of wartime medical care in its time.

Amputation Overcome by Surgical Fever

CASE: Captain Frederick M. Barber, Co. H, 16th Connecticut aged 32 years, was wounded at Antietam, September 17, 1862, by a musket ball, which entered behind the right trochanter major and shattered the trochanters [knobs at the top] and neck [narrow bone at the end] of femur. He was conveyed to the field hospital of the 3d division of the Ninth Corps. His general health was good, and there was but little shock. There was no swelling of the soft parts; the fracture was accessible to exploration, and appeared limited to the epiphysis. The case was one in which excision seemed peculiarly applicable, and, after a consultation of several surgeons of the division, that operation was decided upon. On the morning of September 18th, the patient being anesthetized by chloroform, Surgeon Melancthon Storrs, 8th Connecticut, made a straight incision four inches long, passing through the wound of entrance. The comminuted fragments of the neck and trochanter were extracted, the round ligament divided, the head of the femur removed, and the fractured upper extremity of the shaft was sawn off by the chain saw. The edges of the wound were then approximated by adhesive straps, and simple dressings were applied. Little blood was lost, and the patient rallied promptly from the operation, and appeared quite comfortable during the day. Surgical fever soon set in, however; the patient sank rapidly under the constitutional irritation, and died on September 20, 1862.[13]

Participants can cross back over the bridge and make their way across and up the heights—and "collect their whiskey." They can then proceed to the next stop at the Otto and Sherrick Farm Hospitals (Chapter 3, Stop 2).

STOP 2: OTTO AND SHERRICK FARM HOSPITALS

Figure 3.9 **Hospitals at the Crossroads.** Looking north from the J. Otto Farm side, this photograph of the Sherrick Farm was taken shortly after the Battle of Antietam. The barn in this photo, which was used as a hospital, is no longer on the field of battle. It was a casualty of lightning in 1982 and today only bares the foundation. Source: Library of Congress. *Antietam, Maryland. Sherrick's house, near Burnside bridge.* Gardner, Alexander; September 1862.

Standing at the intersection (**Chapter 3, Stop 2**) in the vicinity of three farms that served as hospitals during the battle, observers can see how the path of least resistance guided the flow of troops. Just like the lowering terrain witnessed earlier at the Roulette Farm (**Chapter 2, Stop 4**), the natural patient flow traveled to this convergence of buildings in the afternoon heat of 17 September 1862.

In 1913, author Clifton Johnson interviewed civilian eyewitnesses to the major battles during the Civil War. Their descriptions of the medical aspects of the fighting in and around Sharpsburg are rich. He made no attempt to censure their testimony nor explain any bias.

Civilian Eyewitnesses to the Battle of Antietam

OTTO FARM FORMER SLAVE AND FOREMAN

Thursday I come home. Befo' I got there I began to see the Johnnies layin' along the road, some wounded and some dead. Men was goin' over the fields gatherin' up the wounded, and they carried a good many to our barn, and they'd pulled unthreshed wheat from the mow and covered the floor for the wounded to lay on. In the barnyard I found a number of Rebels laid in our straw pile and I told 'em the Yankees was comin' to ketch 'em. But they said that was what they wanted—then they'd get a rest.

I was goin' over a stone wall on my way to the house, and there, leanin' against the wall was a wounded Yankee. I asked him when the Rebs left him.

"Last night about twelve o'clock," he said.

I asked him how they'd treated him, and he said: "They found me wounded, and I reckon they did the best they could, but that wasn't much. They didn't have much to do with."

For a while I carried water to the wounded in the barn, and then I went on to town. I wanted to see where my wife was, and after I found she hadn't been hurt I felt considerable better.

A week later the wounded was moved off our place to a camp hospital, and the family come home. The house, as well as the barn, had been used as a hospital, and whatever had been left in it was gone.[14]

FORMER SLAVE WOMAN AT THE TAVERN

"We went out on the street, and there lay a horse with his whole backbone split wide open. The ambulances was comin' into town, and the wounded men in 'em was hollerin', "O Lord! O Lord! O Lord!"

"Poor souls! and the blood was runnin' down thoo the bottom of the wagons. Some of the houses was hospitals, and the doctors was cuttin' off people's legs and arms and throwin' 'em out the do' jest like throwin' out old sticks."[15]

CANAL BOATMAN LIVING ON THE OUTSKIRTS OF SHARPSBURG

Some of the houses in the town were used for hospitals. The doctors would huddle the family all into one little room, or turn 'em out. The house across the way from mine was a hospital, and the family there got what the doctors called camp fever, and some of 'em died.

For three or four days the soldiers was busy out on the battlefield burying the dead. Lots of dead men got pretty strong before they was buried, the weather was so hot; and the stench was terrible—terrible!

On Friday I was engaged in helping drag the dead horses out of town. A farmer with four horses and a black man and myself did that work. We'd hitch a log-chain around a dead horse's neck, and it was all that the four horses could do to drag the carcass over the hills. We burnt what we could on the edge of the town, but fence rails was the only fuel and most of those had been used for campfires. I s'pose we burnt ten or twelve, and we drug nearly as many more out on the farms so as to get the stench away from the town."[16]

Last Push 111

3/3 STOP 3: HIGH POINT

The Red Fez of Courage

Figure 3.10 **Private Charles F. Johnson.** Union Private Charles F. Johnson as a young Zouave patient during the war. Source: *The Long Roll [a US Civil War journal]*. East Aurora, NY: The Roycrofters;1911.

Members of the Union 9th New York Infantry, also known as "Hawkins' Zouaves," crossed the Antietam Creek just south of the bridge by fording over rapids. At the time of the American Civil War, units depicting this group of soldiers were often dressed in characteristic baggy trousers and fez. [The actual Zouaves were a class of light-infantry regiments of the French army that served in North Africa beginning in 1830.] The risks involved with this advancement were soon realized by Private Charles F. Johnson, Company I, as he reached for another bullet cartridge of the 40 he carried as a basic load. The remaining rounds were wet. Aimed rifle shots whizzed about his head like gnawing gnats. Southern cannon ball, grape shot (a cannon charge consisting of small balls), and canisters exploded around his unit. Arm-length railroad iron flew through the air as every harmful matter was shot out toward the advancing blue wall. Other units became commingled within their desperate formations as they clawed to the top of the ridge. One poor fellow tried to catch up with the rest of his regiment. Johnson described the horrible scene.

> A projectile came along with its deafening death-cry, and took him right in his groin, severing his limbs completely from his body. If I could have heard his shriek, it would not have been so horrible, but to see him seize at his limbs, and fall back with a

terrible look of agony, without being able to catch a sound from him-*Oh God! May I never be doomed to witness such a sight again!*"[17]

A Confederate Minie ball found its mark in Johnson's left hip, which left him feeling as if his entire side had been torn away. Bleeding and staying low, Johnson began to move away from the fight. He stated the following:

> I reached the hospital some time before the day's strife was over. In fact, by the time I got there, faint from loss of blood and the exertion, I was of the opinion that the Rebels had gained an immense advantage over us; but it may be, that in my bewildered state, the fierceness of the contest seemed greater at a distance, than when I was actually in the hottest of the fire. I certainly experienced the most terrible depression of spirits imaginable, as the battle-heat gradually wore off, leaving me to realize the full force of what I had been through. And the terrible sights and sounds that met me as I approached the hospital did not tend to relieve my mind. There were already over a hundred of our boys alone, lying on straw and cornstalks, with wounds of all imaginable shapes and sizes. Our tireless Doctor Humphreys [most likely Dr. George Humphries, the same surgeon who treated Private George W. Ard of the 2nd Georgia Infantry] and his assistants were very busy, I can tell you, bandaging, sewing and cutting human flesh. The sights were terrible, but the sounds were more so, although, as a general thing, our boys made light of their wounds. Some one helped me to a rude bed of straw and relieved me of my accouterments [*sic*], and my canteen was refilled, but this comfort was denied us, for I had not been here five minutes, when the Rebels turned on a heavy fire from a neighboring height. Our Doctor ordered all those who could crawl, to start out. "Leave everything, Boys, and go for your lives, they are firing at us." In my hurry I forgot my belt and cartridge box, much to my self-reproach afterward. I have not learned just what damage was done by this beautiful piece of work, but I don't see how every one could escape, the way we were scattered about at the time."[18]

"In making my way to the rear with the other newly-made cripples, I encountered a couple of our cooks who were bringing up freshly cooked meat to the Company, and of course, I wanted some, for like others newly wounded, I felt savagely hungry. I got two chunks of meat, either one as big as my fist, and the way I pitched into them was a caution. I finished them by the time I got to the farm house, to which we had been sent, Miller's Farm House Hospital, and still I craved for more. Nothing was to be had but apples from the orchard, and I was now in such a state I could not move."[19]

Figure 3.11 **Miller Farm Hospital.** A sketch by Union Private Charles F. Johnson depicting the Miller Farm Hospital where he suffered. Source: Johnson, Charles F. *The Long Roll* [a US Civil War journal]. East Aurora, NY: The Roycrofters;1911.

Johnson wrote to himself at the Miller Farm Hospital.

> I don't expect to be sent home, but I do think we have a right to expect the Government to treat us as wounded soldiers deserve to be treated. I should think that we have need of something in the eating line more nourishing than crackers and poor coffee. My breakfast and dinner have been a small piece of pie a friend gave me, for it is absolutely impossible for me to stomach the miserable rations we get here. The talk is that the men receive excellent care in Middletown and Frederick. I do hope they will send us anywhere, so that it is out of this place."[20]

The next leg of the evacuation journey eventually led back to the general hospitals in Frederick, Maryland. Most patients were treated near the battlefield in division-level field hospitals just outside the battlefield. The next stop is the Philip Pry House (Chapter 4, Stop 1).

References

1. Frassanito WA. *Antietam: The Photographic Legacy of America's Bloodiest Day*. New York, NY: Charles Scribner's Sons; 1978:231–235.
2. Banks J. *John Banks' Civil War Blog*. Accessed February 12, 2021. http://john-banks.blogspot.com/. Note: The author is indebted to historian John Banks for his insight and continual research.
3. US Department of the Interior. *Antietam National Battlefield Lesson Two: "One Vast Hospital."* Accessed February 12, 2021. https://www.nps.gov/anti/index.htm. https://www.nps.gov/museum/tmc/Antietam/Antietam_the_aftermath.html
4. Otis GA. *The Medical and Surgical History of the War of the Rebellion (1861–65), Part I, Volume II, Surgical History*. Washington, DC: US Government Printing Office; 1870:130.
5. Robertson JI Jr. A Federal Surgeon at Sharpsburg. *Civil War History*. Vol 6. Issue 2. Kent, Ohio: The Kent University Press; June 1960:141.
6. Robertson JI Jr. A Federal Surgeon at Sharpsburg. *Civil War History*. Vol 6. Issue 2. Kent, Ohio: The Kent University Press; June 1960:142.
7. Robertson JI Jr. A Federal Surgeon at Sharpsburg. *Civil War History*. Vol 6. Issue 2. Kent, Ohio: The Kent University Press; June 1960:142. [Note: The Office of the Surgeon General noted that the idea of missiles flying through the air with near misses were not proven. However, case studies seemed to show a correlation to concussive explosions attributing to internal injuries.] Otis, GA. *The Medical and Surgical History of the War of the Rebellion (1861-1865) Part II, Volume II, Surgical History*. Washington, DC: US Government Printing Office; 1877:28. Otis, GA. *The Medical and Surgical History of the War of the Rebellion Part III, Volume II, Surgical History*. Washington, DC: US Government Printing Office; 1883:706–707.
8. Robertson JI Jr. A Federal Surgeon at Sharpsburg. *Civil War History*. Vol 6. Issue 2. Kent, Ohio: The Kent University Press; June 1960:143.
9. Robertson JI Jr. A Federal Surgeon at Sharpsburg. *Civil War History*. Vol 6. Issue 2. Kent, Ohio: The Kent University Press; June 1960:145–6.
10. Priest JM. *Antietam: The Soldiers' Battle*. Shippensburg, PA: White Mane Publishing Co; 1989:230.
11. Priest JM. *Antietam: The Soldiers' Battle*. Shippensburg, PA: White Mane Publishing Co; 1989:229–234.
12. *The Weekly Banner* (Athens, Georgia); 26 July 1892:6. Note: Special thanks to Laura D. Elliot and her 17 December 2017 post on https://civilwartalk.com/
13. Barnes, JK. *The Medical and Surgical History of the War of the Rebellion Part III, Volume II, Surgical History*. Washington, DC: US Government Printing Office; 1883:92.
14. Johnson C. *Battleground Adventures: Stories of Dwellers on the Scenes of Conflict in Some of the Most Notable Battles of the Civil War*. Boston, MA, and New York, NY: Houghton Mifflin Co; 1915:107–108.
15. Johnson C. *Battleground Adventures: Stories of Dwellers on the Scenes of Conflict in Some of the Most Notable Battles of the Civil War*. Boston, MA, and New York, NY: Houghton Mifflin Co; 1915:111.
16. Johnson C. *Battleground Adventures: Stories of Dwellers on the Scenes of Conflict in Some of the Most Notable Battles of the Civil War*. Boston, MA, and New York, NY: Houghton Mifflin Co; 1915:116–117.

17. Johnson CF. *The Long Roll*, East Aurora, NY: The Roycrofters;1911:190–193.
18. Johnson CF. *The Long Roll*, East Aurora, NY: The Roycrofters;1911:196–197.
19. Johnson CF. *The Long Roll*, East Aurora, NY: The Roycrofters;1911:197.
20. Johnson CF. *The Long Roll*, East Aurora, NY: The Roycrofters;1911:200.

CHAPTER 4

Division-Level Field Hospitals

4/1 STOP 1: PHILIP PRY HOUSE FIELD HOSPITAL MUSEUM

The Philip Pry House Field Hospital Museum located in Keedysville, Maryland, is part of the National Museum of Civil War Medicine in Frederick, Maryland. The museum includes displays of 19th-century medicine and highlights those who administered medical care to their patients. Of particular importance to staff ride participants is the museum's focus on the challenges US Civil War surgeons faced during battle. The med-

Figure 4.1 **The Philip Pry House.** The Philip Pry House as it appeared during the Battle of Antietam where it served as the Headquarters for the Army of the Potomac under Major General George B. McClellan. The original photo taken by Alexander Gardner erroneously labeled the print as "Antietam, Maryland. Gen. Joe Hooker's headquarters." Source: Library of Congress. Gardner, Alexander; September 1862.

ical system developed by Major (Dr) Jonathan Letterman and displayed at the museum became the model for modern battlefield medicine. Before visiting, call the Philip Pry House Field Hospital Museum at (301)416-2395 to confirm the hours or call the main National Museum of Civil War Medicine at (301)695-1864 for more information. There is a small admission fee.

The Philip Pry House (**Chapter 4, Stop 1**), built by Philip Pry in 1844, and the adjacent barn command a view of the terrain from the Antietam Creek to the small town of Sharpsburg, Maryland. For this reason, it was a valuable vantage point for Union Major General George B. McClellan to position his headquarters for some time during the Battle of Antietam on 17 September 1862. As part of the headquarters staff of McClellan's Army of the Potomac, Medical Director Jonathan Letterman would have naturally been positioned here as well to oversee the medical services. Several corps units also passed through this area. Naturally, the flow of the wounded from the early morning fight near D.R. Miller's Cornfield would have also made its way back to this point.

A Journalist's Jaunt

US Army correspondent Charles Carleton Coffin arrived at the battlefield shortly after the fight began. Traveling from Keedysville, just east of this stop, he moved a little further west up the road and described the scene.

> Striking across the fields, I soon came upon the grounds on Hoffman's farm selected for the field-hospitals. Even at that hour of the morning it was an appalling sight. The wounded were lying in rows awaiting their turn at the surgeons, tables. The hospital stewards had a corps of men distributing straw over the field for their comfort. Turning from the scenes of the hospital, I ascended the hill and came upon the men who had been the first to sweep across the Hagerstown pike, past the toll-gate, and into the Dunker Church woods, only to be hurled back by Jackson [Confederate Major General Thomas "Stonewall" Jackson], who had established his line in a strong position behind outcropping limestone ledges. 'There are not many of us left,' was the mournful remark of an officer.

After reviewing the Federal Army's right side, Coffin circled back through the center of the battlefield and crossed back over Antietam Creek.

> Turning from the conflict on the right, I rode down the line, toward the center, forded the Antietam and ascended the hill east of it to the large square mansion of Mr. Pry, where General McClellan had established his headquarters. The general was sitting in an arm-chair in front of the house. His staff were about him; their horses, saddled and bridled, were hitched to the trees and fences. Stakes had been driven in the earth in front of the house, to which were strapped the headquarters telescopes, through which a view of the operations and movements of the two armies could be obtained. It was a commanding situation. The panorama included fully two-thirds of the battlefield, from the woods by the Dunker Church southward to the hills below Sharpsburg.[1]

Illustrator Paints the Scene

Frank H. Schell, of *Frank Leslie's Illustrated Newspaper*, presented earlier in Chapter 1, also passed through Keedysville and described the scene while taking the high ground.

> Hurrying on, I saw near McClellan's headquarters at Pry's farm, on a bare hill beyond it, a group of dismounted officers. I climbed the hilltop, and the group resolved itself into Generals McClellan, Fitz-John Porter, and other officers unknown to me. Aides and couriers were coming and going with fidgety hurry, bringing reports and taking orders. There were moments of impressive silence as, with suppressed mental excitement, all eyes were fired in one direction—toward the distant point to the right.[2]

'How He Escaped with Life is Almost a Mystery'

Dr J. Franklin Dyer served as the regimental surgeon for the Union 19th Massachusetts Infantry and kept a journal throughout his tenure in the Army. He wrote the account below from Valley Mills near Antietam Creek

[Philip Pry House] on 23 September 1862. Many contemporaries referred to the Philip Pry House as Valley Mills, which was the name of area.

> The wounded were fast coming in [Hoffman Hospital just up the road from the Philip Pry House], and in an hour there were five hundred there. Remained there twenty-four hours, when Dr. [G. S.] Palmer, surgeon in chief of the division, having arrived, I got relieved to go and collect together some of the wounded of my own regiment who were at other places. Colonel [Edward] Hinks [19th Massachusetts Infantry Commander], with three other officers, was at Mr. Pry's, Valley Mills. Found him severely wounded, a ball having passed through the right forearm, shattering the radius, and passing through the abdomen, came out on the left of the spine. Surgeons who had examined his wound considered his quite hopeless, and very little had been attempted for his relief. Fixed him up comfortably as possible and then hunted up and got together as many more as possible. They are so scattered however that it is impossible to have those of the same corps, even, together. Some have gone to Hagerstown, others to Frederick, and others leaving for more distant houses or hospitals as soon as they are able.
>
> I cannot say how many thousand wounded are in this vicinity, but every house and barn for miles is filled with them. All the churches and schoolhouses in Sharpsburg are used as hospitals, also those in Keedysville and Boonsboro. At Mr. Pry's house, where I now am, we have five of our wounded officers. Yesterday I got an order from the medical director to accompany some of our wounded Massachusetts officers home, but the colonel was not able to be moved, and I had to send the rest off and remain here. The colonel's wound is a bad one, and it is very difficult to handle him. How he escaped with life is almost a mystery. Dr. [J. N.] Willard is sick here too, not able to take care of the few we now have here, and the army was moved on to Harpers Ferry. There are other wounded officers here; and in the mill barn and other outbuildings are now about one hundred fifty, many of them with

amputated limbs. Some contract surgeons have been sent there, and we shall leave as soon as satisfactory arrangements are made. I do not want to remain here long, the army will soon be on the move, and there is scarcely a medical officer in our brigade with this regiment.[3]

Case Studies from *The Medical and Surgical History of the War of the Rebellion (1861-65)*

The following case study documents the care provided to a wounded soldier through his final record of care. The Office of the Surgeon General collected all field medical reports and produced the largest compilation of wartime medical care in its time.

Commander Puts His Best Foot Forward

CASE: Major-General J. Hooker, U.S. V., was wounded at the battle of Antietam, September 17, 1862. The injury was reported by Assistant Surgeon B. Howard, U. S. A., as follows:

"He was wounded in the right foot by a minie' ball while leading his command, being on horseback at the time, and standing in his stirrups with his weight thrown on his right foot, which was turned outward. The ball struck the inner side of the foot inferiorly to the middle of the scaphoid bone, passing between the first and second layers of the plantar muscles almost transversely across the plantar portion of the foot, and emerging inferiorly to the anterior border of the cuboid bone. The bones of the foot were uninjured. On the morning of September 18th, I was sent by the Medical Director of the Army of the Potomac to attend [to] General Hooker, * * then lying in a farmhouse near the battlefield. Warm-water dressings had been applied previous to my visit. There was no constitutional disturbance, but the foot was hot and inflamed. By means of a syringe I thoroughly washed out the wound with warm water, and finding it most agreeable to the patient, substituted cold- for warm-water dressings. The next day I found the patient very comfortable; the appearance of the foot had greatly improved and the inflammatory symptoms had disappeared. I then ordered a lotion of plumbi [lead salts with sedative properties] instead of cold-water dressings as being more likely to allay any irritation that might arise in the parts. Before the General left that evening for Washington, I advised him to resume the use of tepid water as soon as all tendency to active inflammation should cease. On October 25th, I heard that tetanic symptoms had manifested themselves, but received a letter from the General a few days afterwards stating to the contrary. On November 25th the General, who had returned to duty in the field, requested me to look at his wound, which still troubled him somewhat. I found the newly formed cicatrices somewhat tumefied [scars hardened]; they were painful on pressure, and the General was still unable to mount his horse unaided, though he persisted in being on active duty. On November 30th, I found there had been a steady improvement, and, although the step had not its former elasticity, the wound had left no serious inconvenience behind." General Hooker remained in active service until the close of the war, and was ultimately retired October 15, 1868.[4]

Figure 4.2 **Union Major General Joseph Hooker.** Union Major General Joseph Hooker in uniform with American flag, binoculars, and sword. Source: Library of Congress (Liljenquist family collection). New York, NY: C.D. Fredericks & Co; circa 1861–1868.

Eyes and Ears of the Medical Director

Assistant Surgeon Benjamin Douglas Howard, who personally treated Union Major General Joseph Hooker at the Philip Pry House, was assigned to the Army of the Potomac Headquarters on the staff of Medical Director (Surgeon) Jonathan Letterman. Dr Howard served as the director's "eyes and ears."

Dr Letterman directed medical support under Major General George B. McClellan at the Army of the Potomac Headquarters and described the work performed by Dr Howard.

> I received valuable aid on this occasion from Assistant-Surgeon Howard, U. S. A., who was busily engaged, while the battle was in progress, in riding to different parts of the field, and keeping me informed of the condition of the Medical Department."[5]

[Dr Howard later developed an improved wheeled ambulance (Howard Ambulance). With added suspension, the new design provided much better comfort for patients. Additionally, Dr Howard developed a technique for sealing penetrating (sucking) chest wounds, allowing for respiration. He also created a manual technique for performing artificial respiration that became famous and was used for another 100 years.] Howard described his task as follows:

> On our arrival at Keedysville, another battle was evidently imminent. By order of the medical director, I examined and selected the buildings in town best adapted for hospitals, supervised the necessary preparations, and placed surgeons in charge of the respective hospitals. Surgeon James L. Farley, 84th New York Volunteers, was instructed to act as surgeon in chief of all the hospitals. Two hundred additional ambulances, which I had conducted from Middletown during the night, were in readiness near general headquarters. Hospital tents were entirely wanting. There were on hand no reserve supplies of medicine and hospital stores. Each command had to rely on what they brought with them in their forced marches. Some raw regiments had been hurried forward without medical supplies, and the remainder had, as a rule, an unusually small amount on hand. On September 17th, I was the only remaining medical officer on duty with the director, Surgeon Letterman, the rest of his assistants having been detached to Middletown, Crampton's Gap, and elsewhere, and, according to his instructions, I visited the centre and right of our position, and made the best arrangements I could for the distribution of the wounded. I found several commodious farm-houses, a large barn, and good water at conve-

nient distances, and a large flour mill also, and directed that they should be occupied. * * The wounded were numerous, and it was necessary to lay many of them in the yards contiguous to the houses, that they might be supplied with food from their kitchens, and have their wounds dressed in the open air. . . . * * From the 18th to the 30th of September, the days were very warm; but there usually came a dense and cold fog, which lasted till about nine o'clock the next morning, the fog had a very disagreeable odor, as if impregnated with exhalations from dead bodies on the battlefield. As after most engagements, many of the wounded were destitute of blankets, and it was impracticable to provide them with shelter. This may serve to explain the prevalence of diarrhoea which was greatest about September 24th. Unfortunately, the purveyor was unable to supply the astringents required. I suspected that this diarrhoea had a specific intermittent character, and recommended the use of quinia, combined with Dover's powder [a traditional medicine used for colds and fever], as a substitute for astringents, and this medication had great success. The wounded Confederate prisoners, who were in hospital near Sharpsburg, were generally in an asthenic condition. Tetanus was observed almost exclusively among them, and was seen chiefly amongst those who had marched, before the battle, not less than thirty miles in twenty-four hours. The exposure to which they were subjected during the hot days and cold nights, in which the enemy were effecting their retreat, appeared to have strongly predisposed them to this disease. . . .[6]

Figure 4.3 **Howard Ambulance.** The Howard Ambulance (from the rear) was created by Assistant Surgeon Benjamin Douglas Howard. The tailgate is partially lowered revealing the rollers that supported the cushioned litters. It was originally featured in a publication by the US Sanitary Commission. Source: New York Public Library Digital Collections. Pell, George E; circa 1861–1872.

The Enlisted Club Med

The barn on the right of the Philip Pry House hosted mostly enlisted men who required medical aid after the bloodbath just beyond the Antietam Creek. The house and barn were both built in 1844. Learned men and women of medicine knew from the British experience in the Crimean War (1853–1856) that open air facilities produced better patient outcomes. Dr Letterman wanted to identify as many barns as possible prior to the battle. Per Dr Letterman's instructions, open air refuges were preferable to dank enclosed houses.

> The resources of the country for hospital purposes were ascertained as speedily as possible, and, when an idea was given of the nature of the battle, and the positions to be occupied by our troops, instructions were issued to Medical Directors of Corps to form their hospitals, as nearly as possible, by divisions, and at such a distance in the rear of the line of battle as to be secure from the shot and shell of the enemy to select the houses and

Figure 4.4 **Philip Pry Barn Hospital.** The Philip Pry Barn, located near Antietam Creek, served as a Union hospital as shown in this sketch made at the time of the battle. Source: Johnson, Robert Underwood; Buel, Clarence Clough. *Battle Leaders of the Civil War, Volume II*. New York, NY: Century Co; circa 1884–1888.

barns most easy of access and, when circumstances permitted, to choose barns well provided with hay and straw, as preferable to houses, since they were better ventilated, and enabled Medical officers to attend a greater number of wounded to place the wounded in the open air near the barns, rather than in badly-constructed houses and to have the medical supplies taken to the points indicated.[7]

Of particular interest in the craftsmanship of Philip Pry's Barn is how the mortise and tenon joints are assembled within it. Close inspection of the interior reveals the hand tool marks used to form the shapes. The timber is carefully matched to connected pieces with specific markings indicating their placement.

Figure 4.5 **Mortise & Tenon Joint.** A detailed interior view of a mortise and tenon joint within the Philip Pry Barn. Courtesy of Scott C. Woodard, US Army Medical Department Center of History and Heritage; November 2017.

Barns served as field hospitals and were usually located near the battlefield. It was in these areas that the emerging doctrine of the Ambulance Corps was formed. Inside the Philip Pry House Field Hospital Museum, a reproduction Wheeling, or Rosecrans, ambulance wagon is on display; this model was used early on in the war. Union Major General William S. Rosecrans and Major (Dr) Jonathan Letterman collaborated on its design while they were both stationed at the Department of West Virginia at Wheeling. Lighter than previous four-wheeled ambulances (about 800 pounds), it was easily drawn by two horses. With four elliptical springs, the Wheeling-Rosecrans ambulance wagon could transport 12 ambulatory patients, or two litter and three ambulatory patients, more comfortably than previously refurbished transport wagons. Cushioned seats lined both sides of the interior and could be raised or lowered to adjust to varying patient requirements. Also integral to the design was a 5-gallon water tank under the rear seats; other medical supplies were stored under the front.[8]

Figure 4.6 **Wheeling-Rosecrans.** A reproduction Wheeling, or Rosecrans, ambulance wagon is on display at the Philip Pry House Field Hospital Museum in Keedysville, Maryland. Courtesy of Scott C. Woodard, US Army Medical Department Center of History and Heritage; November 2017.

4/2 STOP 2: SAMUEL PRY MILL HOSPITAL (VALLEY MILLS)

The Samuel Pry Mill (**Chapter 4, Stop 2**) is a private residence and is not available for public access. It was also a private residence in 1862 when it was owned by Philip Pry's brother, Samuel Pry. Its location along the path where Union Corps soldiers traveled toward the D.R. Miller Farm and Cornfield, and the Sunken Road, made it an easy landmark for wounded soldiers returning from battle. The Samuel Pry Mill was an ideal refuge since it was an open air building, close to a road network, and near fresh water. It served as a gristmill for grinding grain. If traveling in a large bus, the guard rail area adjacent to the historical signs along Keedysville Road serves as a good stopping area. Otherwise, the gravel area at the bend of the road can accommodate a passenger van.

Even though 17 September 1862 is remembered as the bloodiest day in American history, the scene to the reader's front where there were once approximately 200 wounded Federal and Confederate soldiers would seem almost calm compared to the devastation just across Antietam Creek. At the time of the battle, medical personnel of the US II Corps, wearing their distinctive green kepi bands and half chevrons, assisted surgeons while transporting patients on Satterlee litters around the mill, turned hospital. A red flag prominently displayed represented a hospital; the more common yellow hospital flag was not standardized until 1864. The much preferred four-wheeled Wheeling-Rosecrans ambulance wagons were available, but the two-wheeled vehicles were still in the inventory and used.[9]

Figure 4.7 **Modern Samuel Pry Grist Mill.** The grist mill and surrounding buildings were selected as hospitals for their ideal characteristics—open air and proximity to the road and water. The Samuel Pry Grist Mill is now a private residence. Courtesy of Scott C. Woodard, US Army Medical Department Center of History and Heritage; November 2017.

4/3 STOP 3: DR OTHO J. SMITH'S FARM (FRENCH'S DIVISION) HOSPITAL

Continuing further along Keedysville Road, the road's name changes to Mansfield Road after crossing Antietam Creek at the Upper Bridge. The approximate location of the Dr Otho J. Smith Farm (**Chapter 4, Stop 3**) in 1862 was on the southern edge of Mansfield Road, just a little over half a mile from the Samuel Pry Mill. There is no parking, so vehicles should carefully pull to the side of the road.

Before the battle, Major (Dr) Jonathan Letterman inspected the various hospitals to see how they could best support the wounded from South Mountain. While making plans to provide medical readiness support on the afternoon of 15 September 1862, Dr Letterman scouted potential hospital locations in the area of operations for the impending battle near Sharpsburg, Maryland. These actions put Dr Letterman, the Medical Director, in danger of becoming a casualty himself.

> In the afternoon of the same day [15 September 1862] we marched through Keedysville, which was subjected to a similar examination [inspecting for hospital sites]. Passing beyond this village we came, about sunset, upon the ground afterwards so widely known as the battlefield of Antietam; and were unpleasantly greeted by the shells from one of the enemy's batteries, which opened upon us as soon as we appeared in sight.[10]

Open barns were available throughout the countryside and were preferred over houses. The number of wounded soldiers soon exceeded the holding capacities of local barns. Guidance was issued for medical personnel to use straw to comfort patients and to provide protection from hostile fire. Dr Letterman explained,

> These directions were generally carried into effect, but the hospitals were not always beyond the reach of the enemy's guns. Very few hospital tents were on hand, owing to the haste with which the army moved from Virginia into Maryland, but fortunately the

Figure 4.8 **Dr Anson Hurd and Barn.** This scene is taken from the vantage point from the other side of the thrashed roof barn pictured in Figure 4.10. The southern mountain range (circled) helps participants identify the location of the site. The sash across Dr Anson Hurd, Regimental Surgeon of the Union 14th Indiana Infantry, pictured among Confederate wounded on the left, indicates he is the officer of the day. Source: Library of Congress. Gardner, Alexander; September 1862.

Figure 4.9 **Dr Otho J. Smith Farm—from the road.** This center image is a modern view of Figures 4.8 and 4.10. The southern mountain range (circled) helps participants identify the location of the site. The authors wish to credit the inspiration to paste the original photographs (Figures 4.8 and 4.10) together from authors William A. Frassanito and John Banks. Frassanito, William A. *Antietam: The Photographic Legacy of America's Bloodiest Day*. New York, NY: Charles Scribner's Sons; 1978:215–223. John Banks' Civil War Blog. Source: Google Maps Street View. Accessed April 2021. https://www.google.com/maps/

Figure 4.10 **Dr Otho J. Smith Farm—from the road.** The Dr Otho J. Smith Farm, located at Keedysville, Maryland, was used as a hospital after the Battle of Antietam. The southern mountain range (circled) helps participants identify the location of the site. Source: Library of Congress. Gardner, Alexander; September 1862.

weather after the battle was so pleasant, that the wounded could be well cared for without them.[11]

Union Brigadier General William French's Division, US II Corps, engaged the Confederates at the Sunken Road (**Chapter 2, Stop 2**). The wounded sought protection and aid in the Roulette Barn (**Chapter 2, Stop 4**), but shelter it was not. As described earlier, because the fighting was much too close, the surgeons began to relocate the soldiers under their care to sanctuaries beyond the rifle balls and artillery shells. Assistant Surgeon Samuel Sexton, Union 8th Ohio Infantry, established a hospital after organizing and caring for the wounded who returned from the front. The medical staff used the barns and improvised shelters made from split rails, which were set up in parallel rows, secured, and thatched with straw. Each makeshift ward held 10 to 15 wounded soldiers. Surgeon Anson Hurd, Union 14th Indiana Infantry, tended to Confederate wounded soldiers near large barns that belonged to Dr Otho J. Smith.[12] When observing the southern mountain range shown in Figures 4.8 through 4.10, viewers are looking towards the Samuel Pry Mill and the Philip Pry House just along Antietam Creek. Bloody Lane and the D.R. Miller Cornfield are approximately 2 miles on the right.

4/4 STOP 4: KEEDYSVILLE HOSPITALS

Traveling back from Dr Otho J. Smith's Barn, staff ride participants can follow Mansfield Road and cross the Upper Bridge; Mansfield Road later changes into Keedysville Road. Follow this road and turn on South Main Street in Keedysville, Maryland. Just as the road today travels from the northern part of the battlefield, this was also the original road that led troops to engagements on the morning of 17 September 1862. The rearward drift of the wounded that afternoon retraced the soldiers' advance that morning. The center area of the village, where the old train depot was located, provides abundant parking for buses (**Chapter 4, Stop 4**). From this center point of the town, participants should not have to walk more than one-third of a mile.

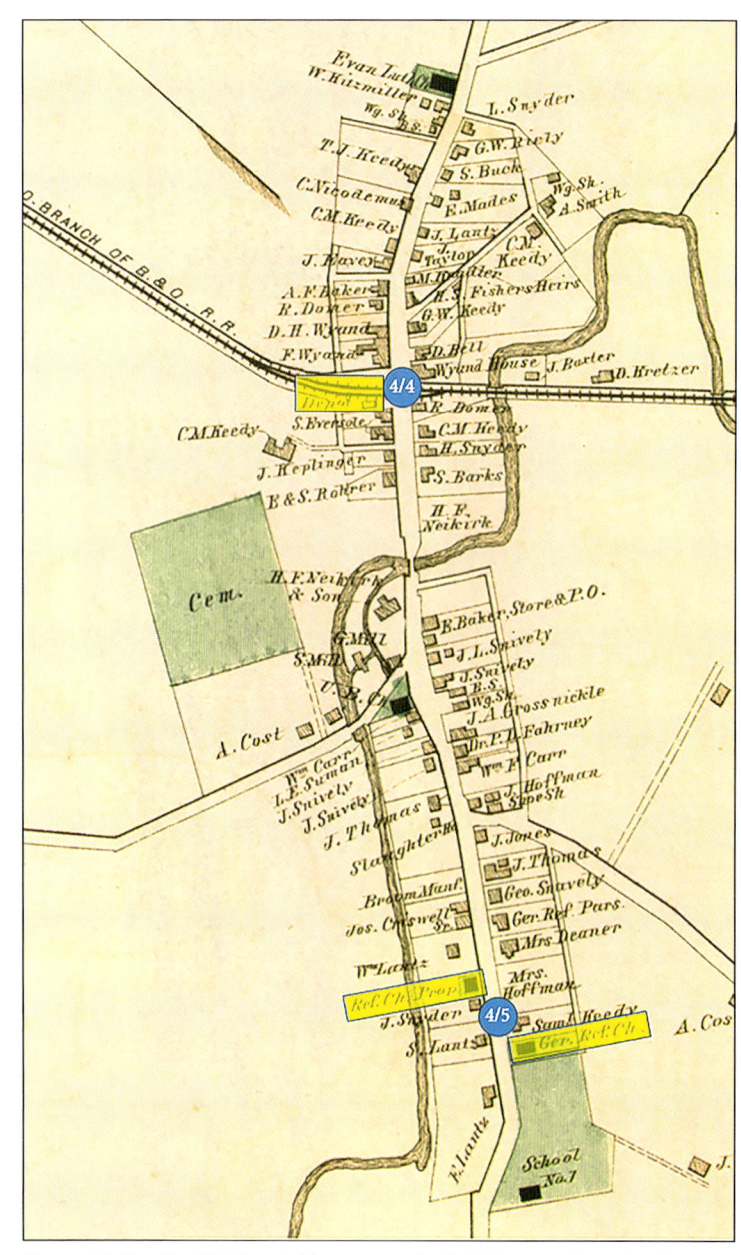

Figure 4.11 **Keedysville Map (with areas marked).** This map of Keedysville, Maryland, was published in 1877 in *An Illustrated Atlas of Washington County, Maryland*. It accurately depicts the village of Keedysville as it appeared in 1862. Source: Johns Hopkins Sheridan Libraries. Philadelphia, PA: Lake, Griffing & Stevenson; 1877.

In describing the action on the right flank of the Federal position, Dr Thomas T. Ellis, a post surgeon from the state of New York, explained,

> Our wounded, as fast as they fell, were carried to the rear, and without delay put into an ambulance, to be carried to the hospitals we had established, and were promptly cared for.

The further "fixed" hospitals in the rear of the formations were reserved for those poor souls unable to journey far from the battlefield.

> The wounded have been all recovered. Those who were able to bear the journey have been sent to Hagerstown, Chambersburg, Harrisburg, and other places. The worst cases we have placed in the houses and barns in the vicinity of the battlefield, which have been fitted up as temporary hospitals.[13]

Figure 4.12 **Fresh-Water Spring.** Fresh water from the Little Antietam Creek and numerous other springs behind the structures off Main Street in Keedysville, Maryland, were essential for bathing, drinking, and laundry. Courtesy of Scott C. Woodard, US Army Medical Department Center of History and Heritage; November 2017.

This Unhappy Village

After his self-evacuation from the battlefield near the Dunker Church (**Chapter 1, Stop 3**) and treatment at his regimental field hospital, Union Colonel Isaac J. Wistar, 71st Pennsylvania Infantry, was placed in an ambulance for his next journey to recovery at Keedysville. Union Private Van R. Willard, of the 3rd Wisconsin Infantry, who was wounded at D.R. Miller's Cornfield (**Chapter 1, Stop 2**), also traveled to Keedysville. Wistar wrote the following:

> Before long an ambulance was brought up and the surgeons decided to send Lieut. Wilson and myself in to [into] the general hospitals at Keedysville. The vehicle jolted horribly over the rough fields, and poor Wilson soon became delirious and died in the ambulance, but I was deposited at a house where Mrs. Wistar had taken up her quarters, to her great relief, as I had been reported dead, since early morning.
>
> The churches of this unhappy village had first been appropriated for the wounded, then successively the houses, shops, yards, and at last the streets, leaving a single track in the middle for the ever-arriving ambulances. . . .
>
> Early on the day of battle, the Keedysville shopkeeper in whose house I found asylum, had crossed the road and entered the opposite field, where he was killed by a stray cannon-shot in the presence of my wife and his own, while trying to see something of the distant battle whose swelling roar already filled the air for many miles around. The house was filled with wounded officers of the 71st [Pennsylvania Infantry], even to the cellar, where lay the adjutant and a captain. After lying here three weeks, my injured artery was pronounced safe for travel, and I was carried in an ambulance to Hagerstown, from whence in a box-car filled with similar convalescents, Mrs. Wistar and I made our slow way via Harrisburg to Philadelphia. Notwithstanding the degrading consciousness of the large space in our lives and memories appropriated by mere physical pleasures, I can never forget the gratification afforded me

while lying in the ambulance at Hagerstown, by Lieutenant Kirby, 1st U. S. Art., who had the patient kindness to hold a cigar for me to smoke, being my first returning dissipation of the kind, as I was still unable to raise either hand to my face.[14]

Later, Brigadier General Wistar became a benefactor of his great uncle's (Dr Casper Wistar) anatomy collection that the doctor developed during his tenure at the University of Pennsylvania from 1792 to 1818. The Wistar Museum, in Philadelphia, Pennsylvania, was renamed the Wistar Institute of Anatomy and Biology in 1892. The Wistar Museum, an anatomical museum, and the original Wistar Institute, an anatomical research institute, were both named after Dr Wistar and were the first of their kind in the United States.[15]

Patients Should Not Require Patience in Waiting for Medical Supplies

Dr Letterman's report addressed the medical supply challenges after the Battle of Antietam and explained how the medic's job continued on after the fighting had ceased. Keedysville, in particular, became the treatment area for many wounded Confederates. In his memoir, Dr Letterman famously quipped, "Humanity teaches that a wounded and prostrate foe is no longer an enemy."[16]

> After night, I visited all the hospitals in Keedysville, and gave such directions as were deemed necessary. The subject of supplies, always a source of serious consideration, was here peculiarly so. The condition of affairs at Monocacy Creek remained as heretofore described, and the action of the railroad was not commensurate with the demands made upon it. The propriety of obtaining the hospital wagons from Alexandria was evident, as these gave a supply for the emergency, and enabled surgeons to attend to the wounded as soon as the battle opened. On the close of the battle, supplies of medicines, stimulants, dressings, and stores were sent for and brought from Frederick in ambulances, and were distributed to the different hospitals as they were needed. The fear of the supplies becoming exhausted, for the difficulty of procuring them was well known, caused uneasiness on the part of some

medical officers, who did not know the efforts that had been made before, and were made during and after the battle, to have enough furnished to supply their wants. I visited, after the battle, every hospital in the rear of our lines, and in no instance did I find any undue suffering for lack of medical supplies. Owing to the difficulty in having them brought from Monocacy Creek, for the first few days, the supplies of some articles became scanty, and in some instances very much so; but they were soon renewed, and, at the temporary depot established in Sharpsburg, shortly after the battle, a sufficient quantity of such articles as were necessary from time to time arrived, and when this temporary depot was afterwards broken up, about the middle of October, a portion of the supplies remained on hand. Not only were the wounded of our own army supplied, but all the Confederate wounded, which fell into our hands, were furnished all the medicines, hospital stores, and dressings that were required for their use."[17]

Figure 4.13 **Zouave Musical Band.** The band of Hawkin's Zouaves play to the sick and wounded in a hospital near Keedysville, Maryland. The actual Zouaves, after whom the Union 9th New York Infantry modeled themselves, were a class of light-infantry regiments of the French army. Schell, F.H. *Frank Leslie's Illustrated Newspaper*. Number 370, Volume XV. New York, NY: Frank Leslie;November 1862:92.

Observations from Members of the Christian and Sanitary Commissions

The Reverend I.O. Sloan, from the US Christian Commission, carried supplies to the battlefield and eventually passed beyond the Samuel Pry Mill property (**Chapter 4, Stop 3**). After settling at the Hoffman House, he described the scene before him.

> At Keedysville several houses were filled with wounded. We halted at the last one, used as a hospital by Sedgwick's Division [Union Major General John Sedgwick]. Every room in the house was filled with wounded, and every spot almost in the yard. This hospital was in charge of Dr. Houston, a good man who was trying to do all he could, but they had as yet received no stores, and were entirely without anything to eat. We gave them what we could spare. From the hills a little beyond here the conflict was plainly visible. All day they were still bringing in wounded to this place. . . . Nearly every farmhouse and barn in all that region was made a hospital; we visited all and left some supplies.[18]

Dr Cornelius R. Agnew, US Sanitary Commission, wrote the following:

> The wounded were mostly clustered about barns, occupying the barnyards, floor[s] and stables; having plenty of good straw, well broken by the power threshing machines. I saw fifteen hundred wounded men lying upon the straw of two old barns within sight of each other [probably the hospitals of French's Division, Sumner's Corps]. Indeed there was not a barn or farmhouse, or store, or church, or schoolhouse, between Boonesboro, Keedysville, Sharpsburg, and Smoketown that was not gorged with wounded—rebel and Union. Even the corncribs, and in many instances the cow stables, and in one place the mangers were filled. Several thousand lie in the open air upon straw, and all are receiving the kind services of the farmers' families and of the surgeons. But everything in the way of medical supplies was deficient; poor fellows with broken and lacerated thighs had to

be carried out of barns into the open fields to answer to a call of nature. . . . Concentrated foods are also scanty; in fact everything wanting that wounded men need, except a place to lie down, and the attention of personally devoted surgeons.[19]

Observations from Members of the Medical Services

Union Surgeon Charles Fitz Henry Campbell, of the 1st Division, US I Corps, assumed the duties of administration in addition to medical care.

> The wounded from this greatest battle of modern times were scattered in buildings everywhere contiguous to the field. . . . I was on duty at Keedysville, a few miles removed from the field, for two days, engaged in perfecting arrangements for the reception of the wounded, and superintending the reception and distribution of supplies.[20]

Union Assistant Surgeon Charles Carroll Gray, Army of the Potomac Headquarters, fought death with food and medicine. Of note is his comment at the end about providing specimens to the US Army Medical Museum.

> I arrived at Sharpsburg, Maryland, September 19, 1862, and was assigned by Medical Director Letterman to the charge of a hospital in Keedysville, and, a few days afterward, the army having advanced, was made a sort of issuing commissary for the wounded in and about Keedysville. My instructions were to hire transportation from the citizens, and to draw and deliver all rations, preventing the use of ambulances for this purpose. The cattle furnished me being of poor quality, I exchanged a considerable portion of the meat with citizens, receiving therefor [therefore] milk, eggs, vegetables, etc. At the time of my arrival at this battlefield, I consider that the wounded were as well cared for as it was possible they could be. Abundance of supplies soon poured in from a variety of sources. Of the strength of the army at the time of action, I know nothing. Medical and hospital stores were plentiful.

The wounded were attended to at sundry points, varying from half a mile to two miles from the field. They were not exposed to rain; the nights, however, were chilly, and there was considerable suffering on this account. The wounded were mostly removed in ambulances. Almost all the wounds I saw were from conoidal balls, and a large number, I should suppose, received at short range. So many of the wounded as were deemed capable of bearing an ambulance transportation of eighteen miles, were sent to Frederick, Maryland. The remainder, especially such as had undergone capital operations, were collected in two permanent field hospitals, where it was contemplated to afford every advantage of our best general hospitals. It appears to me that this arrangement was wise, and must have saved lives.

Amputations were the rule, and in general promised well. I saw but two or three excisions. There were, however, a considerable number of tetanus cases. All under my observation resulted in death. Chloroform was the anaesthetic generally used. I observed no bad results therefrom. I operated but twice: first, an amputation of the leg at the point of election, which was successful; second laryngotomy for oedema glottidis, which was unsuccessful. The larynx and trachea in the latter case were sent to the Army Medical Museum.[21]

4/5 STOP 5: GERMAN REFORMED CHURCH HOSPITAL (MOUNT VERNON REFORMED CHURCH)

The Bloody Work of Amputation Commenced

The German Reformed Church Hospital at Keedysville, Maryland, is located just over one-third of a mile south of the old train depot area and served as one of the larger sites in this quaint little town for hosting the wounded. The original German Reformed Church was built in 1852 on land donated by Samuel Cost. Because of a foundation failure, the original church building, used as a hospital after the battle, was replaced by the new Mount Vernon Reformed Church in 1892. This church was made possible by the efforts of leading families in the community: the Keedy, Pry, and Cost families. Witnesses to the heroic medicine performed within those walls included carpeting, portieres (doorway drapes), pews, and a chandelier which were all part of the original 1852 church. All of these items were included in the rebuilt edifice in 1892. It is unknown whether the pits dug for amputated limbs contributed to the eroding foundation, but the new chapel was built over a mass grave for soldiers who died of their wounds or sicknesses while they were there.[22]

Union Private George A. Allen, 76th New York Infantry, was detailed as a hospital steward and assigned to the German Reformed Church Hospital in Keedysville following the battle. His experience was published by *The Antietam Wavelet*, a local Keedysville paper, on 29 March 1890.

> The principal hospital was established in the brick church near the upper end of the town. Boards were laid on top of the seats, then straw and blankets, and most of the bad cases of wounded were taken to this, the headquarters. Comrades with wounds of all conceivable shapes were brought in and placed side by side as they could lay, and the bloody work of amputation commenced. The surgeons, myself, and a corps of nurses with sleeves rolled up, worked with tender care and anxiety to relieve the pain and save the lives of all we could.
>
> A pit was dug just under the window at the back of the church and soon as a limb was amputated I would take it to the window and drop it outside into the pit. The arms, legs, feet and hands that were dropped into that hole would amount to several hundred pounds.
>
> On one occasion I had to fish out a hand for its former owner, as he insisted that it was all cramped up and hurt him.
>
> Every morning those who died during the night were taken out and buried in a trench, usually without ceremony. But one morning, as I was directing some North Carolina conscripts that had surrendered, and were set to work digging graves, an officer,

Figure 4.14 **German Reformed Church.** The German Reformed Church (left) in Keedysville, Maryland, was used as a Union hospital. Source: Johnson RU, Buel CC. *Battle Leaders of the Civil War, Volume II*. New York, NY: Century Co; 1884–1888:635.

Figure 4.15 **Mount Vernon Reformed Church.** Rebuilt in 1892, the former Union hospital is now the Mount Vernon Reformed Church (below). Notice the V Corps badge displayed on the roof apex. Courtesy of Scott C. Woodard, US Army Medical Department Center of History and Heritage; November 2017.

Figure 4.16 **Interior of Church.** Witnesses to the heroic medicinal efforts within the church walls (above) included carpeting, portieres (doorway drapes), pews, and a chandelier from the original 1852 church; they were included in the rebuilt edifice in 1892. Source: Reed, Paula. Survey #WA-II-1074, Street Address 64 South Main Street. Keedysville, MD. Preservation Associates, Inc.; November 1993. Courtesy of the Maryland Historical Trust.

I think a lieutenant, staggered up drunk and prepared to give the remains a send-off; and had I not caught him by the arm he would have fallen in to [into] the trench. With the wave of a hand and hiccough, he said, "Let the Father take the spirit that gave it." He then braced up and staggered off with a self-assurance that he had performed a solemn duty.

Most of the wounded got it into their heads that no one but myself could dress their wounds, change their bandages, etc., so I had to do the most of it, and was kept busy for several weeks, night and day. For three weeks I never realized that I had not slept at all. I was in several battles and on many hard marches, but nothing ever wore me as that did. At night I would drop down upon the sofa in the pulpit, but no sooner had I closed my eyes than some one of the many amputees would call me to change a bandage or something. When a patient gets it into his head that you can handle his stump more carefully than any one else, you're elected. . . . Dr. [Bernard] Vanderkief[t] was the boss in taking off a limb. He could snatch a leg or arm off quicker than you could say "jack Robinson."

The little brick church at the southern end of Keedysville, still stands and is still being used by the Reformed denomination. Mrs. Howard Bartiner, J. A. Miller's daughter, told me the people of Keedysville still tell of the hospital and its patients. But I wonder if, when the congregation of this little church sits in their pews on Sunday morning, their thoughts ever do go back to the brave boys who lay wounded and dying within those red brick walls.[23]

Stone School

Dr Elisha Harris, US Sanitary Commission of New York, inspected the hospitals surrounding the Antietam battlefield. In his *Report on Field Hospitals indicated on Map of Battlefied of Antietam,* he described the hospitals in Keedysville that were located in various buildings:

Figure 4.17 **Stone School.** Described as the "Stone School," this home was used as one of the many hospitals in Keedysville, Maryland. Courtesy of Scott C. Woodard, US Army Medical Department Center of History and Heritage; November 2017.

"Brick store; Barn; Old Mill; Old Brown House; Stone School House."[24]

The Stone School House is located just across the street and a little north of the German Reformed Church. The house is now a private residential dwelling; it originally served as a school and church building. Before the war, the Reverend Robert Douglas preached in this very building. Having grown up nearby, his son, Henry Kyd Douglas, was instrumental in guiding the Confederate forces in the Maryland Campaign of 1862. He later penned the memoir, *I Rode with Stonewall*.[25]

Leaving the Hospitals and Moving East

Union Private Franklin Flint Thompson joined the 2nd Michigan Infantry and served on temporary duty as a nurse during and after the Battle of Antietam. Thompson shared an incredible story of treating a soldier on the battlefield who confessed in his last breaths to actually being a female. Maintaining the secrecy, Thompson buried the dead soldier without notifying the command of her true identity. Later, Thompson went into detail in describing a scene he witnessed of a pompous government clerk desperately trying to get back to Washington, DC, after touring the battlefield. The clerk, however, was stymied in his homeward journey because all of the trains were full of wounded soldiers.

> After the cars moved off there they stood gazing after it in the most disconsolate manner. Said one, "I came out here by invitation of the Secretary of War, and now I must return on foot, or remain here." One of the soldiers contemptuously surveyed him from head to foot, as he stood there with kid gloves, white bosom, standing collar, etc., in all the glory and finery of a brainless fop, starched up for display. "Well," said the soldier, "we don't know any such individual as the Secretary of War out here, but I guess we can find you something to do; perhaps you would take a fancy to one of these muskets," laying his hand on a pile beside him.[26]

His memoir closes with the possible motivation of thought behind the description of this encounter.

> But when I look around and see the streets crowded with strong, healthy young men who ought to be foremost in the ranks of their country's defenders, I am not only ashamed, but I am indignant!"[27]

The two scenes may or may not have actually happened. Much of Thompson's memoir of the war may have been apocryphal and actually revealed his own feelings at the time. Thompson was a real soldier who served almost two years in the US Army, but when his memoir of the war was published in 1865, the author was Sarah Emma Edmonds. The short boyish soldier was actually a woman who remained undetected throughout her service in the Army.

Case Studies from *The Medical and Surgical History of the War of the Rebellion (1861-65)*

The following case study documents the care provided to a wounded soldier through his final record of care. The Office of the Surgeon General collected all field medical reports and produced the largest compilation of wartime medical care in its time.

Unable to Shoulder the Burden

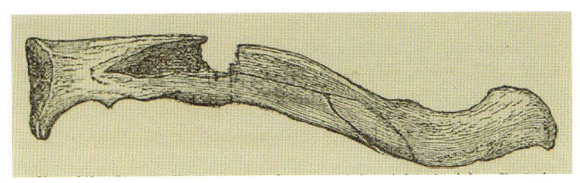

Figure 4.18 **Right Clavicle**. A longitudinal gunshot fracture of the right clavicle (posterior view). This specimen was donated to the US Army Medical Museum. Otis GA. *The Medical and Surgical History of the War of the Rebellion (1861–65), Part I, Volume II, Surgical History, Chapter V, Wound and Injuries of the Chest*. Washington, DC: US Government Printing Office; 1870.

CASE: An unknown soldier was wounded at Antietam, September 17th, 1862, by a conoidal musket ball, at short range. The missile entered at the junction of the inner third with outer two-thirds of the right collar-bone, made a clean perforation in the anterior wall of the bone, and largely splintered the posterior portion, and emerged above the right scapula. The wounded man was carried to the field hospital at Keedysville. On admission, he was speechless, and in a fainting condition from loss of blood. The track of the wound was plugged with lint saturated with the solution of the persulphate of iron [solution used to control bleeding]. The usual restoratives were cautiously administered, and the strictest quiet enjoined. On September 19th, 1862, a deluging hemorrhage occurred, and the patient almost immediately expired. It was found that a spicula of the clavicle had transfixed the left subclavian. The artery was not preserved. The clavicle, represented in [**Figure 4.18**], was presented to the Museum by Assistant Surgeon S. Storrow, U.S.A.[28]

Before departing the now quiet village of Keedysville, ponder Dr Thomas T. Ellis' words in early October 1862.

"Our hospitals at Keedysville and Sharpsburg have today been cleared of all our wounded, but the rebel wounded still remain."[29]

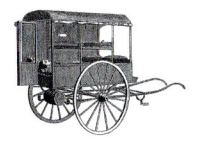

The next phase of studying the medical aspects of the battle require traveling down the evacuation route to Frederick via Middletown, Maryland. Participants can access the original National Road, Alternate US Highway 40 (Old National Pike) toward Middleton, for the drive into the town.

References

1. Johnson RU, Buel CC. *Battle Leaders of the Civil War, Volume II*. New York, NY: Century Co; 1884–1888:683.
2. Schell FH. Sketching Under Fire at Antietam. *McClure's Magazine*. New York, NY, and London, England: S. S. McClure Co;February 1904:418.
3. Chesson MB. J. *Franklin Dyer's The Journal of a Civil War Surgeon*. Lincoln, NE: University of Nebraska Press; 2003:40–41.
4. Otis GA. *The Medical and Surgical History of the War of the Rebellion, Part III, Volume II, Surgical History*. Washington, DC: US Government Printing Office; 1883:60.
5. Letterman J. *Medical Recollections of the Army of the Potomac*. New York: NY: D. Appleton & Co; 1866:40.
6. Woodward JJ. *The Medical and Surgical History of the Rebellion (1861-65), Part I, Volume I, Medical History, Appendix to Part I of The Medical and Surgical History of the Rebellion, containing Reports of Medical Directors and other Documents*. Washington, DC: US Government Printing Office; 1870:104–105.
7. Letterman J. *Medical Recollections of the Army of the Potomac*. New York, NY: D. Appleton & Co; 1866:39.
8. Otis GA. *The Medical and Surgical History of the War of the Rebellion, Part III, Volume II Surgical History*. Washington, DC: US Government Printing Office; 1883:949–950.
9. "Island of Mercy: The Pry Mill at Antietam." Certificate of authenticity distributed with each original signed print issued by the National Museum of Civil War Medicine in Frederick, Maryland.
10. Letterman J. *Medical Recollections of the Army of the Potomac*. New York, NY: D. Appleton & Co; 1866:38–39.
11. Letterman J. *Medical Recollections of the Army of the Potomac*. New York, NY: D. Appleton & Co; 1866:39.
12. Johnson RU, Buel CC. *Battle Leaders of the Civil War, Volume II*. New York, NY: Century Co; 1884–1888:672.
13. Ellis TT. *Leaves from the Diary of an Army Surgeon; or Incidents of Field Camp, and Hospital Life*. New York, NY: John Bradburn, publisher; 1863:280–281,301.
    Note: Dr Thomas T. Ellis previously served in the British army in South Africa as a post surgeon and was wounded during the Boer Wars.
14. Wistar IJ. *Autobiography of Isaac Jones Wistar (1827–1905) in Two Volumes, Volume II*. Philadelphia, PA: The Wistar Institute of Anatomy and Biology; 1914:68–69.
15. Wistar IJ. *Autobiography of Isaac Jones Wistar (1827–1905) in Two Volumes, Volume II*. Philadelphia, PA: The Wistar Institute of Anatomy and Biology; 1914:163–167.
16. Letterman J. *Medical Recollections of the Army of the Potomac*. New York, NY: D. Appleton & Co; 1866:46.
17. Woodward JJ. *The Medical and Surgical History of the War of the Rebellion (1861-65), Part I, Volume I, Medical History, Appendix to Part I, Containing Reports of Medical Directors, and Other Documents*. Washington, DC: US Government Printing Office; 1870:97.
18. Duncan LC. *The Medical Department of the United States Army in the Civil War*.

Carlisle Barracks, PA: Medical Field Service School; 1931:Chapter V,26–27. Note: Duncan originally published this work in *The Military Surgeon, Volume XXXII, Number 5*, in May 1913.

19. Duncan LC. *The Medical Department of the United States Army in the Civil War*. Carlisle Barracks, PA: Medical Field Service School; 1931:Chapter V,27–28.

20. Woodward JJ. *The Medical and Surgical History of the Rebellion (1861-65), Part I, Volume I, Medical History, Appendix to Part I of The Medical and Surgical History of the Rebellion, containing Reports of Medical Directors and other Documents*. Washington, DC: US Government Printing Office; 1870:106.

21. Woodward JJ. *The Medical and Surgical History of the Rebellion (1861-65), Part I, Volume I, Medical History, Appendix to Part I of The Medical and Surgical History of the Rebellion, containing Reports of Medical Directors and other Documents*. Washington, DC: US Government Printing Office; 1870:105–106.

22. Kerns J, Thomson D. *Historic Keedysville Walking Tour*. Mercersburg, PA: Mercersburg Printing; 2018.

23. Allen G. Scenes in the Hospital at Keedysville [Maryland]. *The Antietam Wavelet*; 29 March 1890.

24. Harris E. Report on Field Hospitals indicated on Map of Battlefield of Antietam. The original report resides at the Maryland Center of History and Culture (formerly the Maryland Historical Society), Prints and Photographs Department (Map Collection). Baltimore, MD;1862.

25. Reed P. Survey #WA-II-10-78, Street Address 55 South Main Street (Architectural survey file used to document historic properties). Hagerstown, MD: Preservation Associates, Inc; November 1993.

26. Edmonds S, Emma E. *Nurse and Spy in the Union Army*. Hartford, CT: W. S. Williams & Co; 1865:279.

27. Edmonds S, Emma E. *Nurse and Spy in the Union Army*. Hartford, CT: W. S. Williams & Co; 1865:384.

28. Ellis TT. *Leaves from the Diary of an Army Surgeon; or Incidents of Field Camp, and Hospital Life*. New York, NY: John Bradburn, publisher; 1863:307.

29. Barnes JK. *The Medical and Surgical History of the War of the Rebellion (1861-65), On Special Wounds and Injuries*. Washington, DC: US Government Printing Office; 1870:522.

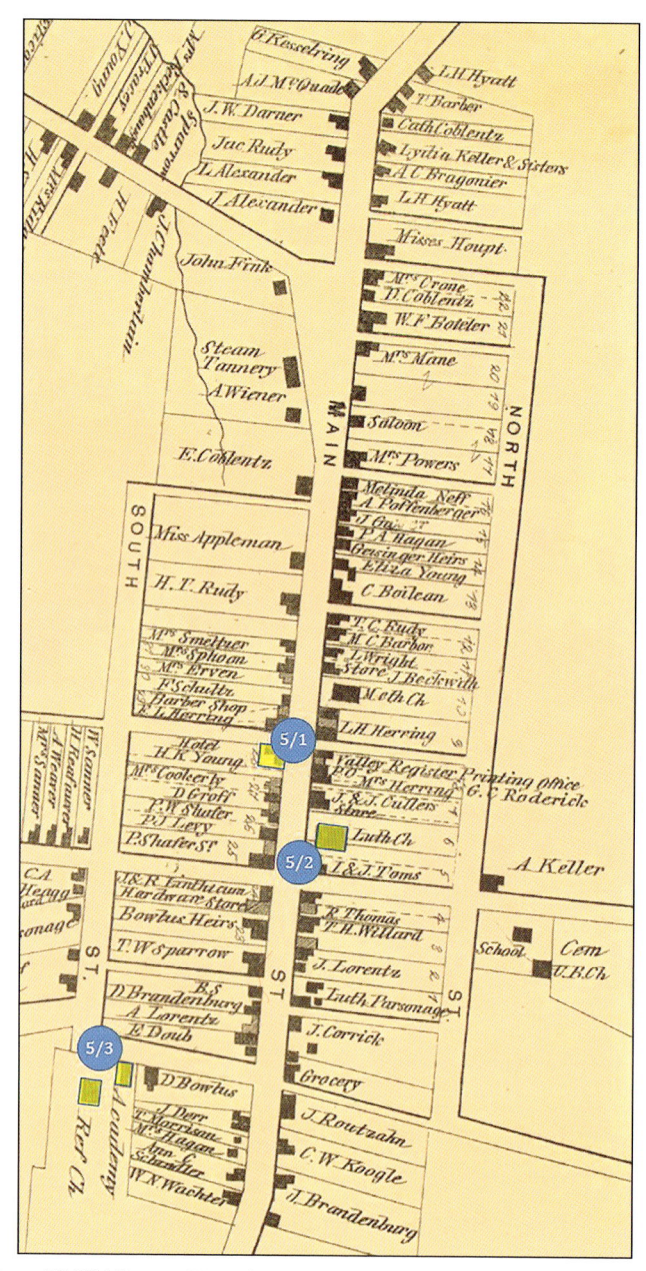

Figure 5.1 **Middletown Map.** This map of Middletown, Maryland, was published in 1873 in the *Atlas of Frederick County, Maryland* publication and roughly depicts the town as it appeared in 1862. Source: Johns Hopkins Sheridan Libraries (Maryland state, county, and Baltimore city atlases); 1873.

CHAPTER 5

Corps and Evacuation Hospitals

5/1 STOP 1: MAIN SUPPLY ROUTE

It is a welcomed rest traveling from Keedysville, Maryland, toward Frederick. The heart of Middletown, at the intersection of Main Street and Elm Street (**Chapter 5, Stop 1**), provides a convenient parking area for participants, just as it served as a waypoint for wounded soldiers along the evacuation route many years ago. Modern asphalt and traffic obstruct the view, but 19th-century witnesses saw a well-maintained turnpike connecting civilization from east to west.

Establishment of Corps Hospitals

Before 1862, the original National Road, Alternate US Highway 40 (Old National Pike), later named US Route 40 Alternate, was a well-established road linking Baltimore, Maryland, to St. Louis, Missouri. Just as it made a logical path for supplies and men to the battles of South Mountain and Antietam, this main supply route also served as the evacuation road back to Frederick. As Major (Dr) Jonathan Letterman surveyed the resources and topography, he began to establish a predetermined sanctuary for the wounded.

> ❝ ... The village of Middletown, about four miles in rear of the scene of action, was thoroughly examined before the battle began, to ascertain its adaptability for the care of the wounded.

Churches and other buildings were taken, as far as was considered necessary, and yet causing as little inconvenience as possible to the citizens residing there. Houses and barns, the latter large and commodious, were selected in the most sheltered places, on the right and left of the field, by the medical directors of the corps engaged, where the wounded were first received, whence they

Figure 5.2 **1862 Middletown Sketch.** Union troops marching through Middletown, Maryland, en route to South Mountain on 14 September 1862. Source: Library of Congress. Waud, Alfred R; 1862.

Figure 5.3 **Modern Main Street.** Modern view of Main Street in Middletown, Maryland, looking toward the Zion Lutheran Church. Notice the road dimensions have not changed since the 1862 sketch. Source: Google Maps. Accessed May 15, 2021. https://maps.google.com

were removed to Middletown, the Confederate wounded as well as our own. The battle lasted until some time after dark, and as soon as the firing ceased I returned to Middletown and visited all the hospitals and gave such directions as were necessary for the better care of the wounded.[1]

Lines of Evacuation

Not every corps unit received ambulances in time for commanders to make effective use of their people and other assets. Nonetheless, the battlefield was cleared of casualties by the evening of 18 September 1862. Dr Letterman explained,

> I have already mentioned that the ambulances had been left at Fort Monroe, when the troops embarked, and that no system existed, except in the corps which belonged to the Army of the Potomac, while at Harrison's Landing. A portion of the ambulances of some of the corps arrived just prior to the battle; a large number had been distributed in other corps, but were yet unorganized, and was not expected that they would prove as efficient as was desired. Notwithstanding, the wounded were brought from the field on our right before two o'clock on the following day. The ambulance train of the second Corps was more fully equipped, and did most excellent service, under the charge of Captain J. M. Garland, who labored diligently, and with great care, until all his wounded were removed. The troops on the left were those among whom no well organized ambulance system existed; but here, owing to the exertions of the medical officers, the wounded were removed by the evening of September 18th. When we consider the duration and magnitude of the engagement, and the obstinacy with which it was contested, it is a matter of congratulation to speak of the expeditious and careful manner in which the wounded were removed from the field. Compiled from the most reliable sources at my command, the number of wounded amounted to eight thousand three hundred and fifty. This number is not entirely accurate, as many who were slightly wounded were attended to, of whose cases no record could under the circum-

stances, be taken. The removal of so large a body of wounded was no small task. The journey to Frederick in ambulances was tedious and tiresome, and often painful to wounded men. It was necessary that they should halt at Middletown for food, and to take rest; that food should always be provided at this place at the proper time, and for the proper number; that the hospitals at Frederick should not be overcrowded; that the ambulances should not arrive too soon for the trains of cars at the depot at Frederick, the bridge over Monocacy Creek having been rebuilt; and that the ambulance horses should not be broken down by the constant labor required of them. With rare exceptions, this was accomplished, and all the wounded whose safety would not be jeopardized by the journey, were sent carefully and comfortably away.[2]

Correspondent's Witness

The Valley Register, a weekly newspaper in Middletown, Maryland, published the following article on 26 September 1862 from a New York reporter traveling through town the previous week.

> Surgeons with hands, arms and garments covered with blood, are busy amputating limbs, extracting balls and bandaging wounds of every nature in every part of the body. Rebel soldiers in great numbers lie among our own and receive the same attention.
>
> Keedysville, Boonsboro, Middletown, and I presume Frederick, are rapidly being filled with the wounded from the battles of Sunday [South Mountain] and Wednesday [Antietam].
>
> The inhabitants in all these villages are laboring night and day to relieve the dying and the suffering. A more Christian people, in the practical significance of that word, I never saw. Every private dwelling is filled with the wounded. Carpets are torn up, costly furniture is removed, comfortable mattresses spread upon the floor awaiting the arrival of the ambulances.

And much of this preparation for the wounded is without one word from the medical directors in regard to it. In the pleasant village of Middletown, especially, I have seen nothing in the hospitals in Washington that indicated so much thoughtfulness and devotion. All the ladies in the village are spending night and day with the wounded.[3]

5/2 STOP 2: ZION LUTHERAN CHURCH HOSPITAL

"May he send you help from the sanctuary and grant you support from Zion." Psalm 20:2

The Zion Lutheran Church (**Chapter 5, Stop 2**) is located just east and across the street from the previous stop in Middletown, Maryland, and was the largest structure in Middletown at that time. It began its service as a US Army hospital when blood stained the South Mountain on 14 September 1862 before the great convergence near Sharpsburg, Maryland, on the 17th. The sanctuary was only 3 years old when it welcomed wounded soldiers from the Battle of South Mountain, just five miles west. Pews were removed and replaced with cots for the wounded. The wounded from the Antietam battlefield gorged the already full emergency hospital. Today's carpets conceal the blood stains on the church floor.[4]

The US Christian Commission's Reverend I.O. Sloan (introduced earlier in Chapter 4) passed through and described the scene.

> We arrived in Middletown on the morning of the seventeenth, here we found wounded men coming in from the battlefield, some with fingers shot off, arms broken, wounded in the head, covered with blood. The little church in the main street [Zion Lutheran Church] was already filled with our wounded, as also some of the houses opposite. . . . As we hurried along to where the two armies were engaged we frequently had to stop and give of our supplies to the wounded men whom we met in ambulances, and who lined the road, hobbling along as best they could, to find some temporary hospital.[5]

5/3 STOP 3: GERMAN REFORMED CHURCH HOSPITAL (CHRIST REFORMED UNITED CHURCH OF CHRIST)

Figure 5.4 **German Reformed Church (front view).** Both the adjacent school and sanctuary of the German Reformed Church were used as vital hospitals along the evacuation route back to Fredericksburg, Maryland. Courtesy of Scott C. Woodard, US Army Medical Department Center of History and Heritage; November 2017.

The German Reformed Church (**Chapter 5, Stop 3**) began taking in patients beginning with the Battle of South Mountain on 14 September 1862. The church and the adjacent Reformed Academy served as vital sanctuaries for the flow of broken and dying soldiers coming in from the mountainous melee. From the southern parking lot of the church, participants can view South Mountain from the ground-level. Union Major General George B. McClellan climbed this impressive steeple to view the Battle of South Mountain as it unfolded.

Figure 5.5 **Side view of the German Reformed Church looking at South Mountain.** From the viewpoint of the participant, South Mountain can easily be seen along the horizon while looking west. Courtesy of Scott C. Woodard, US Army Medical Department Center of History and Heritage; November 2017.

The steeple of the German Reformed Church offered an excellent view of Army movements, and it also served as an excellent signal station. Time consuming carrier dispatches were made obsolete with the ability to send communications using the "wigwag" system. This visual signal system, which used a flag during the day and a torch by night, was developed by former Assistant Surgeon Albert J. Myer while he served as a US Army surgeon accompanying troops in Texas in the 1850s. By using his earlier experience as a telegraph operator and his expertise in nonverbal communication from his earlier work with the deaf, Dr Myer developed a new messaging system that was eventually adopted by the Army. Dr Myer later became the first chief of the US Army Signal Corps in June 1860.[6]

References

1. Woodward JJ. *The Medical and Surgical History of the War of the Rebellion (1861–65), Part I, Volume I, Medical History, Appendix to Part I, Containing Reports of Medical Directors, and Other Documents.* Washington, DC: US Government Printing Office; 1870:96–97.

2. Woodward JJ. *The Medical and Surgical History of the War of the Rebellion (1861–65), Part I, Volume I Medical History, Appendix to Part I, Containing Reports of Medical Directors, and Other Documents.* Washington, DC: US Government Printing Office; 1870:97–98.

3. Crossroads of War: Maryland and the Border in the Civil War. Accessed June 4, 2020. http://www.crossroadsofwar.org/research/newspapers/?id=836

4. Zion Lutheran Church 275th Anniversary. Accessed June 4, 2020. https://www.zionmiddletown.org/about-us/275th-weekly-history-pages.html

5. Duncan LC. *The Medical Department of the United States Army in the Civil War.* Carlisle Barracks, PA: Medical Field Service School; 1931:Chapter V,26.

6. Raines RR. *Getting the Message Through: A Branch History of the U.S. Army Signal Corps.* Washington, DC: US Army Center of Military History; 1996:5–8.

❝ Figuratively speaking this city is one vast hospital, and yet hundreds of poor fellows continue to arrive. . . . ❞

—*Philadelphia Inquirer*, 25 September 1862

Figure 6.1 **Frederick Map.** This map was originally published in 1860 and provides locations for the stops highlighting the National Museum of Civil War History and general hospitals in Frederick, Maryland. Source: Library of Congress, Geography and Map Division. Bond, Isaac. *Map of Frederick County, Md.: accurately drawn from correct instrumental surveys of all county roads*; 1860.

CHAPTER 6
Frederick General Hospitals and the National Museum of Civil War Medicine

The start and end point for this portion of the staff ride is the National Museum of Civil War Medicine at 48 East Patrick Street, Frederick, Maryland 21705, (301) 695-1864.

There is a charge for admission, and it is recommended that groups call ahead for special pricing and assistance for parking large vans and buses. Street parking is required for large vehicles and must be coordinated prior to the visit. Privately owned vehicles may park behind the museum in the public parking garage.

This walking tour of selected buildings, used as military hospitals during the war, represents only a fraction of those that are still standing. Each general hospital in the city normally consisted of several buildings, churches or both. The hospitals highlighted in this chapter are not a complete list of the named general hospitals. The entire counterclockwise circuitous route is just over 1 mile in length.

Establishment of General Hospitals

Union Major (Dr) Jonathan Letterman, Medical Director of the Army of Potomac, purposely planned for hospitals to be located within the Army of the Potomac's area of operations. In addition to the division-level hospitals (in Maryland) visited earlier in Keedysville and the corps-level facilities in Middletown, general hospitals were established further in the rear in Frederick.

> In addition to the hospitals in the city [Frederick], two large camps of hospital tents were formed on the outskirts of the city, capable of containing one thousand beds each. One hospital had been established in Frederick some months before our arrival; but at that time it was filled chiefly with Confederate sick and wounded, who had been left there. All the available buildings in this city, six in number, were taken for hospitals. . . . These were fitted up with great rapidity, the buildings selected and prepared; beds, beddings, dressings, stores, food, cooking arrangements made; surgeons, stewards, cooks and nurses detailed, and sent for.[1]

Dr Letterman's memoir speaks to the challenges in planning for continued care away from the point of injury on the battlefield and the enormity of the wounded in Frederick.

> The removal of such a large number of wounded from the field to the General Hospitals was an arduous undertaking. The railroad from Greencastle, Penn., could not be depended upon. That from Harper's Ferry was in no better condition; it was therefore necessary that the wounded should be sent in ambulances to Frederick, for transportation to Baltimore, Washington, and elsewhere. It was imperative that the trains should leave at the proper hours, no one interfering with another; that they should halt at Middletown, where food and rest, with such surgical aid as might be required, could be given to the wounded; that food

Note on general hospitals: Normally, once patients were stable, they were transported from corps area hospitals to general hospitals. As seen in Frederick, Maryland, it was common to name the hospital complex by a numbered system designation (General Hospital Number 1, General Hospital Number 2, etc). It was not unusual to move the entire staff to another building but maintain the same naming mechanism.

should always be prepared at this village at the proper time, for the proper number; that the hospitals in Frederick should not be overcrowded, and the ambulances should arrive at the railroad depot in Frederick at the required time to meet the Baltimore trains. With rare exceptions this was accomplished, and all the wounded, whose lives would not be jeopardized, were sent carefully away. Surgeon J. J. Milhau, U. S. A., was placed in charge of the Frederick Hospitals, to which were added two large camps of hospital tents, each capable of accommodating one thousand patients. On the 30th of October the hospitals of this city contained over five thousand patients, attended by sixty-two surgeons, fifteen medical cadets, twenty-two hospital stewards, five hundred and thirty-nine nurses, and one hundred and twenty-seven cooks. No one, who saw them after they were established, can form any conception of the labor required to put them in operation. Every thing had to be, as it were, created; the place itself supplied nothing but some buildings. These hospitals, as I found from personal inspection, were in excellent order, and the wounded attentively and skilfully [sic] treated.[2]

Ambulance Training Pays for Wagon Trains and Railroad Trains

Even though some corps units had very little time to gather necessary resources before the conflict, leaders in the Medical Department throughout the Army of the Potomac benefited from obtaining ambulances before the battle began. Of particular importance for Dr Letterman was the rapid fielding of the ambulance systems. Clear guidance and an overall plan for the evacuation and treatment of the wounded were staffed and sent to each Corps Medical Director.

Assistant Surgeon John Theodore Reily (Sixth Army Corps Artillery), noted, "Having received orders to convey the wounded of the corps to Frederick, as soon as practicable, I started, a few days after the battle, with two trains of fifty ambulances, and removed them to Frederick without accident. When I arrived there, the medical director ordered me to take charge of a railroad train filled with wounded, going to Philadelphia the following morning."[3]

Doctors and Wounded Converge on Frederick

Acting Assistant Surgeon James Henry Peabody, along with about 70 other doctors in Washington, DC, were ordered to Frederick to assuage the flow of wounded coming from Sharpsburg. He was initially charged with overseeing the US Hotel Hospital and then General Hospital Number 2. Dr Peabody recalled the following:

> I was ordered to report to Medical Director Letterman, Army of the Potomac, for temporary duty in the field. . . . We were, however, detained in Frederick City by order of the Medical Director, and immediately set to work in the care of the wounded, who were by this time arriving by the thousand. The greater number of medical officers having been sent on toward the field, the work for those left in Frederick was almost incessant for a few days. . . . After the battle of Antietam, most of the wounded were hurried on to Frederick, and from thence, those but slightly wounded, after being allowed a night's rest, were transferred to Washington and Baltimore. The hospitals in Frederick were densely crowded after the battle, and every available building used for hospital purposes. Some of these buildings were but poorly ventilated and ill adapted for this purpose; they were given up as speedily as possible. For the first five or six days, owing to crowding, it was almost impossible to keep the sick and wounded supplied with food and other necessities. . . . The greatest inconvenience to which the wounded were exposed was in consequence of our not having an adequate number of beds to accommodate the thousands who were pouring in; and those but slightly wounded had to lie on the floor or ground, as they preferred, until the day after their arrival, when they would be

From the National Museum of Civil War Medicine, participants can travel east on Patrick Street for one block and then turn left and go north on Carroll Street for two blocks. After that, they can stop at the intersection of Carroll and 2nd Streets. General Hospital Number 5 (Chapter 6, Stop 1) is on the right.

transferred to Washington or Baltimore. I have counted as high as twelve hundred thus transferred in one train of cars. This crowding only continued for a few days, after which we had ample supplies and accommodations for those left. Thousands of those wounded in the upper extremities at the battle of Antietam walked in to Frederick City, some eighteen or twenty miles, all the ambulances being constantly busy in the removal of the more severely wounded.[4]

6/1 STOP 1: GENERAL HOSPITAL NUMBER 5, VISITATION ACADEMY

General Hospital Number 5 (**Chapter 6, Stop 1**) consisted of two buildings and was used by the Union Army. The two buildings belonged to the Roman Catholic Church and included the Roman Catholic Novitiate (a Jesuit Seminary) and Visitation Academy (a Sisters of Charity school for girls). Of the two, only the Visitation Academy building remains. The hospital, collectively, could serve up to 517 patients. It included a staff of seven surgeons and assistant surgeons (total), two medical cadets, two hospital stewards, 70 male nurses, and 14 cooks. The Brothers and Fathers of the Jesuit Seminary provided additional aid. These staff numbers do not account for female nurses or volunteers. As a Catholic, it was fitting that Surgeon Henry S. Hewit, a veteran of the Mexican–American War, was assigned as the Surgeon in Charge of the complex. It is noteworthy to draw conclusions from Dr Hewit's writings during his assignment. They foreshadowed the formation of the US Army's Sanitary Corps and Medical Administrative Corps branches another half century away.

Dr Hewit wrote that surgeons in charge should "be relieved from the detail of administration and the care and responsibility of public property beyond their books and instruments" and those tasks should be given to "a class of non-professional persons." Additionally, he said they "ought to have the rank of Lieutenant at least and might be appointed from the most deserving and competent Hospital Stewards."[5]

'O God, Be Thou with Me'

Assistant Surgeon Cyrus Bacon Jr was assigned to the hospital system in Frederick following his appointment as a Regular Army officer. Dr Bacon previously served as a surgeon in the Union 7th Michigan Infantry. His initial assignment was at General Hospital Number 5. Here, he described his fourth day on duty:

> **September 25th [1862]**
> I get tired easily, and have to work very hard. We have operations every day, some days several. This is hard work. I put a man in my wards under chloroform for the extracting of a bullet. It is interesting to see how busy the brothers are around the sick. Dressing wounds, counting their beads over the sick beds, saying prayers. I do not doubt but some men die in Catholic hands because we have no Protestants as active [as they].[6]

Practicing the learned knowledge that patient outcomes were much better in open spaces, Dr Bacon wrote, as officer of the day, "I have to have much care used that the windows of the wards are not closed. Should they be suppurating wounds [they] would soon poison the air. Pure air is worth more than medicine." Later, Dr Bacon was directed to serve in one of two large tent hospital encampments in Frederick-Camp A. His patients were mostly Confederates and he wrote of removing lead bullets. "In a ball which I removed from the muscular tissue of [*the*] thigh of one of my patients here, was found a hammering out flat, quite as if it had been upon the anvil under the hammer. It did not appear to have struck the bone, yet was in the leg deep and spread out flat and thin almost as if intentional."[7]

Case Studies from *The Medical and Surgical History of the War of the Rebellion (1861-65)*

The following case study documents the care provided to a wounded soldier through his final record of care. The Office of the Surgeon General collected all field medical reports and produced the largest compilation of wartime medical care in its time.

Tarheel Donates Leg to the Union

CASE: Private B. Fields, Co. C, 27th North Carolina, aged 21 years, was wounded in the right leg and taken prisoner at Antietam, September 17, 1862.

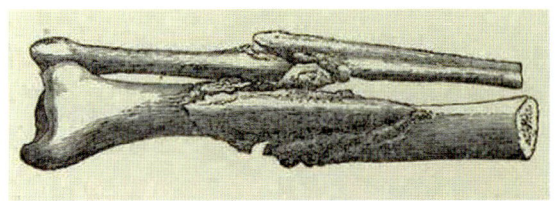

Figure 6.2 **Tibia and Fibula.** Ununited comminuted fracture in the right tibia and fibula. Source: Otis, George A. *The Medical and Surgical History of the War of the Rebellion Part III, Volume II, Surgical History.* Washington, DC: US Government Printing Office; 1883:537–538.

He entered hospital No. 5, at Frederick, ten weeks afterwards, where Surgeon H. S. Hewit, U. S. V., recorded the following description of the injury: A gunshot wound of the tibia and fibula by a minié ball; continuity of bone entirely destroyed; leg very much swollen, offensive, and filled with pus. Flap amputation at the upper third of the leg was performed on November 28th by Acting Assistant Surgeon A. V. Cherbonnier. The stump was closed with three stitches and wet strips of muslin. Reaction was very satisfactory; patient cheerful. On the next day the dressing was removed and some adhesive strips were applied, the stump looking well but there being some pain, for which one grain of morphia [morphine] was administered.

On November 30th the patient looked and felt well, having slept nearly all night; appetite good; wound suppurating well; stump dressed with lint and cold water and well supported with a roller. Secondary hæmorrhage occurred on December 1st, when the popliteal artery [knee area] was ligated by Surgeon Hewit. One week afterwards the ligature had not yet come away, but there was hæmorrhage from the wound, the cause of which was not very obvious. The tourniquet was then applied to the femoral and the wound filled with charpie [lint] saturated with solution of persulphate of iron [solution used to control bleeding].

By December 11th the wound in the popliteal space was suppurating: stump healthy looking and dressed with wet strips of muslin; patient pale in appearance, and ordered to take three grains of quinine and five grains of tartrate of iron and potassa [emetic properties] every four hours. Early on the morning of December 10th another tolerably severe hæmorrhage took place, the popliteal ligature not yet having come away; continued the tourniquet to the femoral and prescribed quinine and iron, cod-liver oil, and egg-nog. Two days later the tibia was found to protrude; tourniquet still kept applied, though a little slackened. On December 18th there was another hæmorrhage, the loss of blood being about six ounces, controlled by tightening the tourniquet.

By December 20th reaction was slowly established, though the patient's face was still bloodless; appetite tolerably good; treatment continued. An attack of diarrhœa had been promptly arrested by the administration of rhubarb powder, ipecacuanha [emetic properties], and opium. December 24th, patient rallying, though the stump still gives evidence of a low state of vital powers. Several days afterwards the stump looked healthier, and on December 29th, when the patient was transferred to hospital No. 1, he was in good spirits and good granulations were springing up.

Assistant Surgeon R. F. Weir, U. S. A., in charge of the latter hospital, recorded the subsequent progress of the case as follows: "The stump continued to look and suppurate well, the patient having but very little pain and his bowels remaining in good condition; appetite improving. For three days previous to January 26th he complained of great tenderness along the inner side of the thigh, where, on examination, fluctuation was detected, and where an incision was made which was followed by the evacuation of ten ounces of pus, the sinus extending under the deep muscles of the posterior and outer aspect. Injections of hydrochloric acid and laudanum [opium and alcohol] diluted in water were then used, and by January 27th the ulcer of the stump had nearly cicatrized [healed with a scar] and the cavity of the abscess was rapidly filling up; patient improving in general health and able to sit up most of the time. About two weeks later the patient was in very good condition going about on crutches, the stump cicatrizing rapidly and looking healthy. On February 28th the patient was discharged from hospital treatment, his general condition having improved very rapidly and the ulcer of the stump having healed."

Several days afterwards the patient was paroled and sent south. The Confederate hospital records show that the man was admitted, on March 18th, to the General Hospital at Petersburg, and that he was permitted to leave for his home, on furlough, April 6, 1863. The greater portion of the amputated bones of the injured leg, contributed to the Museum [National Museum of Civil War Medicine] by Dr. Cherbonnier, are represented in the annexed wood-cut [Figure 6.2], showing an ununited comminuted fracture of both bones in the lower part, followed by an effusion of a large amount of callus, fragments being attached to the fibula, while the tibia is carious within.[8]

Case Studies from *The Medical and Surgical History of the War of the Rebellion (1861-65)*

The following case study documents the care provided to a wounded soldier through his final record of care. The Office of the Surgeon General collected all field medical reports and produced the largest compilation of wartime medical care in its time.

Only 5 Percent

CASE: Private I. Ostheimer, Co. F, 66th New York, aged 31 years, was wounded in the right leg, at Antietam, September 16, 1862, and admitted to a Second Corps field hospital. Surgeon C. S. Wood, 66th New York, reported: "A minié ball struck the tibia anteriorly about four inches above the malleoli [area around the ankle], shattering the bone. The fibula and bones of the foot were uninjured. I performed resection of the lower four inches of the tibia, removing the bone from the articulation, after which splints were applied to the leg; case sent to General Hospital."

Acting Assistant Surgeon A. V. Cherbonnier recorded the patient's admission to hospital No. 5, at Frederick, where a large sequestrum [dead bone or tissue] was removed on November 22d, also that a large portion of the diseased tibia was resected on December 2d. About two weeks afterwards the wound assumed a healthy appearance, and by December 28th it was filling up with healthy granulations, the patient being cheerful and feeling well. On the next day he was transferred to hospital No. 1, where he remained until the following June, when he was sent to Patterson Park (Convalescent) Hospital at Baltimore. On August 10, 1863, the patient was returned to his command for duty, and while in the field he again came under the notice of Surgeon Wood, who made the following supplementary report: "The man laid in hospital for seven months before he could move his leg. Being possessed of a good constitution and no untoward symptoms appearing, at the end of that time he began to walk a little, and now—eleven months after the reception of the injury—he has just been returned to duty. On examination a large cicatrix [healed scar] is found, the bone having refilled its entire length and uniting with the astragalus [bone of ankle joint] with a moveable articulation. Although less than usual, the bone is not quite so long as its fellow, allowing the foot to turn slightly inward. Otherwise he has a very useful limb. He cannot endure hard in arches as well as formerly, and probably never will. Still the case is one of interest, as not one in twenty saves his limb after the receipt of a like injury."

According to information from the Adjutant General, U. S. A., this man has been reported as "missing in action" since the battle of Todd's Tavern, May 8, 1864.[9]

From the Visitation Academy in Frederick, participants can walk west for one block and then turn left and walk south at the intersection of 2nd Street and Maxwell Avenue. After one block, the group can turn right and head west from the intersection of Maxwell Avenue and Church Street. General Hospital Number 4 (Chapter 6, Stop 2) was located here. A Lutheran church with the Winchester Seminary is across the street on the right.

6/2 STOP 2: GENERAL HOSPITAL NUMBER 4

Lutheran Church (Evangelical Lutheran Church)
Winchester Seminary/Frederick Female Seminary (Winchester Hall)

The hospital complex in Frederick, Maryland, consisting of the Methodist Episcopal Church, the Winchester Seminary, and the Evangelical Lutheran Church were all designated as General Hospital Number 4 (**Chapter 6, Stop 2**). Today, only two of the complex's buildings remain. During the war, the patient population reached 279 with a staff consisting of six surgeons and assistant surgeons (total), two medical cadets, three hospital stewards, 28 male nurses, and 10 cooks. As relayed from the previous stop, this staff count does not include the female nurses and volunteers.[10]

Evangelical Lutheran Church

Figure 6.3 **Lutheran Church.** The Lutheran Church, or the Evangelical Lutheran Church as it is named today, was a landmark for miles around Frederick, Maryland, and served as an Army hospital following the fighting at South Mountain and Antietam. Courtesy of Scott C. Woodard, US Army Medical Department Center of History and Heritage; November 2017.

The two Norman Gothic Revival style towers are the distinguishing features of the Evangelical Lutheran Church. It was these twin towers that inspired the opening of John Greenleaf Whittier's famous poem about notable Unionist Barbara Fritchie (or Frietchie): "Up from the meadows rich with corn, Clear in the cool September morn. The clustered spires of Frederick stand Green-walled by the hills of Maryland."

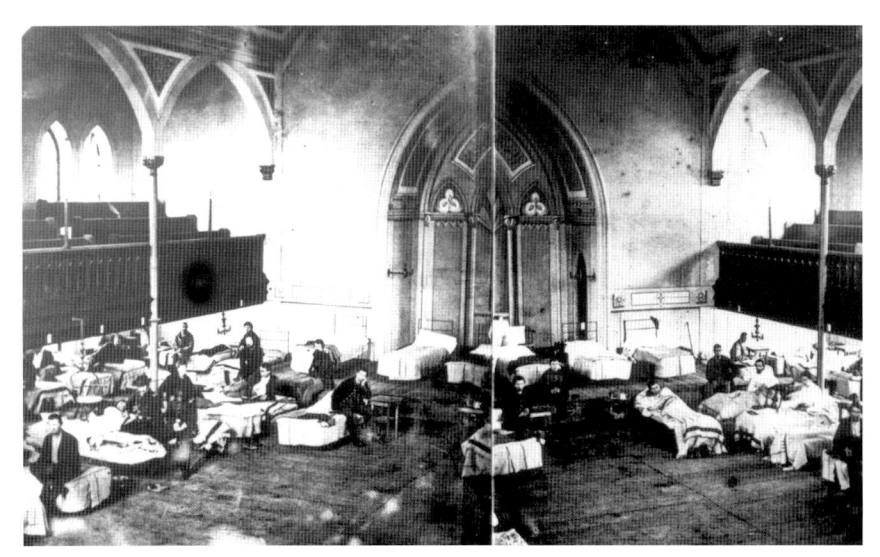

Figure 6.4 **Lutheran Church—inside front.** A composite of two photographs taken from the back of the sanctuary. This is the inside view of the Lutheran Church when it served as a hospital following the battles of South Mountain and Antietam. Source: Evangelical Lutheran Church, Frederick, MD. Accessed June, 15, 2020. https://www.twinspires.org/history

Figure 6.5 **Lutheran Church—inside back.** A composite of two photographs taken from the front of the sanctuary. This is the inside view of the Lutheran Church when it served as a hospital following the battles of South Mountain and Antietam. Source: Evangelical Lutheran Church, Frederick, MD. Accessed June, 15, 2020. https://www.twinspires.org/history

Inside the western tower are two bells purchased in 1771 that still call the church congregation to worship today. The wounded soldiers from the battles of South Mountain and Antietam who recovered there heard those bells as well. In an effort to protect the pews from potential damage from the church's conversion to a hospital, a platform was built on top, creating a false floor. Medical staff members used the adjacent room while guards were stationed in the balconies. Members of the congregation filled critical staff requirements by feeding patients and staff members, rolling bandages, and caring for the wounded. These services, primarily filled by the women of the church, also included writing letters to soldiers far from home.[11]

Winchester Seminary/Frederick Female Seminary

The Greek Revival style columns that stand today are a physical reminder of the temple of higher learning that was the Hiram Winchester's school

Figure 6.6 **Frederick Female Seminary.** The Frederick Female Seminary at Frederick City, Maryland. This undated print describing the private girls' school was created before the US Civil War. Hiram Winchester, Seminary President, sold the school after the war concluded. Source: Library of Congress (Popular Graphic Arts). No date recorded.

Figure 6.7 **Winchester Seminary.** The Winchester Seminary, Winchester Hall as it is named today, now serves as the county seat for Frederick County, Maryland. Wounded soldiers from the South Mountain and Antietam battles were treated here. Courtesy of Scott C. Woodard, US Army Medical Department Center of History and Heritage; November 2017.

for girls in Frederick. When it opened in 1845, the Frederick Female Seminary (previous name) provided a private academic setting for girls. The 1850 school catalog advertised the school as an "an institution where all may have their daughters instructed in the branches of science and literature adapted to the female mind." However, this ideal soon conflicted with reality in September 1862 when the building was used as a Union hospital. The school president negotiated an agreement where the girls maintained their abode and classes on one side (western wing) of the building, while the soldiers were nursed back to health on the other (eastern and older wing). Many families pulled their girls from the school during this time. The old "Winchester Seminary," as it was called, is now Winchester Hall and today serves as a Frederick County government office.[12]

Mary's Beau

Many wounded from the Union 3rd Wisconsin Infantry recovered in Frederick after the Antietam battle. Several were patients in General Hospital Number 4. A month after nurse Clara Barton's discovery of the wounded Union "soldier" Mary Galloway, nurse Barton visited a Frederick hospital room that had about 100 recovering soldiers. One patient caught her attention. In the patient's delirium, caused by a gangrenous arm that was to be amputated, he kept calling out for a "Mary." Barton read the card at his headboard, "H. B., 3d Wis." Having previously treated Mary, Barton had now met both members of the couple. Barton was able to finally reunite them on the eve of Harry Barnard's amputation which he survived.[13]

Case Studies from *The Medical and Surgical History of the War of the Rebellion (1861-65)*

The following case study documents the care provided to a wounded soldier through his final record of care. The Office of the Surgeon General collected all field medical reports and produced the largest compilation of wartime medical care in its time.

An Incredible Recovery

CASE: Private George H. Bowes, 8th Illinois Cavalry, in a skirmish, September 13, 1862, was shot in the abdomen. Captain J. D. Ludlam, 8th Illinois Cavalry, certifies that this man "was shot in a cavalry skirmish, by the enemy, near Middletown, Maryland, and left on the field. I afterward sent an ambulance and brought him in. I did not think he would live through the night. I saw him when shot, and I was commanding the squadron."

Surgeon C. Hard, 8th Illinois Cavalry, does not refer to the case on his monthly report. As most of the wounded of the battles of South Mountain and Antietam were taken to Frederick the search for the patient was directed there, and it was found that Assistant Surgeon H. A. DuBois, in charge of Hospital No. 4, records, that Bowes entered that hospital on September 19th, with a shot wound believed to involve the intestines. The particulars of the progress and treatment of the case are not recorded. On January 5, 1863, the patient was transferred to the hospital at Camp B, Frederick, where Assistant Surgeon T. G. MacKenzie recorded the case without any details.

On March 9th, the patient was transferred to Jarvis Hospital, Baltimore, and came under the charge of Assistant Surgeon D. C. Peters, in whose language a more detailed history may be given:

> George H. Bowes, aged 19, a private in the 8th Illinois Cavalry, was transferred from Frederick, Maryland, to this hospital, March 7, 1863. The patient states that the day previous to the battle of South Mountain his regiment was in the advance, skirmishing with the enemy, when he became engaged in a hand to hand encounter with a rebel horseman. The man fired several shots at him with his revolver, one of which took effect in his abdomen. The ball entered the abdomen about two inches above the umbilicus [navel] and one inch to the left of the linea alba [white line; midline of abdomen], traversed backward and slightly upward, and made its exit just beneath the tenth rib, at a point that is about two and one-half inches from the spinous process of its vertebra. The wound immediately placed him *hors de combat* [out of action], and he commenced to vomit blood, and it at the same time poured from his nostrils. The free hæmorrhage caused syncope [fainting], which temporarily arrested it, but, at spells for the following seven days, he had a series of these hæmorrhages. He further states that after receiving the wound he had bloody passages from his bowels, which gave him intense pain, and continued for about the same length of time. There was but

a small amount of blood that escaped from the wounds. The surgeon who examined him on the field informed him that the ball had passed through his body. The injury was followed by acute inflammation, as he complains of having suffered much pain and tenderness in the whole abdomen, and says he had fever. He was confined to his bed, undergoing active treatment, for several weeks. Whenever he received fluids or solids into his stomach, he states that, for a period of two months, a part of the half-digested material would escape from the anterior wound and soil the dressings. From his system not receiving proper nutrition, he became very weak and emaciated; but finally the wounds closed, and since then he has regained his health rapidly. The healthy action of the primae viae [main passages–bowels] is again fully established, but, owing to contractions formed in the healing of the track of the wound, he is bent forward, and cannot by any force straighten himself. The treatment at present is directed toward overcoming these contractions. Remarks: Cases of recovery from gunshot wounds of the abdomen are by no means uncommon; but recovery from wounds of the stomach (and there is every probability this comes under that category) and other abdominal viscera [intestines] are exceptional to the general rule."

Private Bowes was discharged from hospital and from the military service April 2, 1863. His pension claim was admitted November 24, 1863, on his captain's certificate, already quoted, and a certificate of disability by Dr. Peters, which was substantially an extract from the foregoing report. The disability was rated as total. No further particulars are given by any pension examining surgeon. The pensioner was last paid in September, 1872, his condition being described as unchanged.[14]

From General Hospital Number 4, participants can travel west on Church Street for one block to the intersection of Church and Market Streets.

Of interest along the route is Kemp Hall on the southeast corner of Church and Market Streets. This building served as the temporary location for the General Assembly for the state of Maryland in April 1861. Annapolis, the capital of Maryland, was occupied by Federal troops; Maryland state senators and delegates met there on 27 April 1861. Legislators later met at Kemp Hall during the summer, initially to discuss the state's reaction to Federal occupation and the use of railroads to invade Virginia. Ultimately, the decision on whether to secede was deemed outside of the legislature's authority and the topic was put back for a vote when the body reconvened in September 1861. Any doubt to Maryland's willingness to secede from the United States was answered by the state legislature's inability to form a quorum. The members of the legislature who were thought to favor secession were jailed by Federal soldiers when the Maryland General Assembly met on 17 September 1861.[15]

Just beyond the intersection along Church Street are the Old German Reformed Church (now known as Trinity Chapel, Evangelical Reformed Church) on the left and the German Reformed Church (now known as the Evangelical Reformed Church) on the right. These two church buildings were incorporated into the expansive General Hospital Number 3, which is at the next stop. At the time of the US Civil War, the same congregation owned both sites and Kemp Hall—just as it does today. The German Reformed Church basement was used as a hospital during the Federal hospital expansion, so the sanctuary was still active throughout the Union Army's occupation. Introduced in Chapter 4, Henry Kid Douglas wrote about Confederate Major General Thomas "Stonewall" Jackson's visit in his memoir, *I Rode with Stonewall*. On a Sunday evening, 7 September 1862, Jackson and a few staff officers attended a worship service at the German Reformed Church where Douglas' pro-Union friend, the Reverend Dr Daniel Zacharias, preached. Douglas recalled the following:

> As usual, the General went to sleep at the beginning of the sermon, and a very sound slumber it was. His cap, which he held in his hand on his lap, dropped to the floor, and his head dropped upon his breast, the prayer of the Congregation did not awaken him, and only the voice of the choir and the deep tones of the organ broke his sleep. The Doctor [Zacharias] was afterwards credited with much loyalty and courage because he prayed for the President of the United States in the presence of Stonewall Jackson. Well, the General didn't hear it, but if he had I've no doubt he would have joined in it heartily.[16]

From the Old German Reformed Church and the German Reformed Church, participants can walk west toward the next intersection at Church and Court Streets. Turning north at that intersection, they should then go one block to the intersection of Court and 2nd Streets. As the participants continue west along 2nd Street, they will see the Frederick Presbyterian Church a half block down on the right, which is the first location of General Hospital Number 3 (Chapter 6, Stop 3). Note: Figure 6.1 shows Stop 3 in the vicinity of the three different church buildings of General Hospital Number 3.

6/3 STOP 3: GENERAL HOSPITAL NUMBER 3

Presbyterian Church (Frederick Presbyterian Church)
New Protestant Episcopal Church (All Saints' Episcopal)
Old Protestant Episcopal Church (All Saints' Episcopal)

General Hospital Number 3 (**Chapter 6, Stop 3**), with its extensive staff and buildings, served as one of the larger hospital complexes in Frederick, Maryland. This review will focus on the hospital's three main buildings: the Presbyterian Church, the New Protestant Episcopal Church, and the Old Protestant Episcopal Church. However, several additional properties were included in the original hospital center—Coppersmith Hall, Methodist

Protestant Church, and Frederick Academy (Bonsall's Academy). These are no longer standing. Surviving today, and introduced in the walk en route to this stop, are the remaining hosts of General Hospital Number 3: the German Reformed Church (Evangelical Reformed Church) and the Old German Reformed Church (Trinity Chapel, Evangelical Reformed Church). Based upon the official patient census, this facility was probably used for convalescing soldiers since the turnover rate was so low. The staff included nine surgeons and assistant surgeons (total), two hospital stewards, 65 male nurses, and 14 cooks. Just as indicated in previous staff models, the female nursing staff and volunteers were not documented in official records.[17]

Presbyterian Church (Frederick Presbyterian Church)

During this time, the church played a large role in educating local residents. Frederick Academy was located across the street from the Presbyterian Church at the time. The academy was founded by a part-time pastor of the church with pastors serving as principals. Hiram Winchester of Winchester Seminary (**Chapter 6, Stop 2**) served as a trustee of the church. Like the Lutheran Church (**Chapter 6, Stop 2**), a floor covering was placed on the pews to protect them while the church was used as a hospital.[18]

Figure 6.8 **Presbyterian Church.** The Presbyterian Church, named the Frederick Presbyterian Church today, served as a Union hospital after the battles of Stone Mountain and Antietam in September 1862. Courtesy of Scott C. Woodard, US Army Medical Department Center of History and Heritage; November 2017.

Our Wounded Fez in Frederick

Readers may recall Private Charles F. Johnson, of the 9th New York Infantry (also known as "Hawkins' Zouaves") from **Chapter 3, Stop 3**. Johnson described his 1862 journey from the battlefield hospital at the Miller house near the battlefield.

> **Frederick City, September 29th**
>
> Yesterday morning, Sunday, a train of ambulances came to Miller's Farm Hospital and took away a hundred of those who could travel. You can easily guess that I was glad to be one of the number. We got started about nine o'clock and took the same road that the army used in their advance from this place. . . . Maryland is free from the tread of Rebel hosts, and the Nation rejoices in a great victory, as it mourns over the countless dead. I am returning with many others, helpless, to the little city of Frederick, seeking tenderness and care. The same city we left long before daylight two weeks ago, with musket in hand, hastening to the scene of action then in progress in the mountains, we will enter tonight, weary, sick and hopeless, praying but for rest and the return of strength.

Figure 6.9 **Presbyterian Church—inside.** The scene inside the Presbyterian Church as drawn by patient Charles Johnson. Johnson served as a Union Private with the 9th New York Infantry. Johnson, Charles F. *The Long Roll*. East Aurora, NY: The Roycrofters;1911.

Figure 6.10
President Lincoln's Visit. *The President's visit to the Army of the Potomac—arrival at the station at Frederick.* New York, NY: New York Public Library Digital Collections (The Miriam and Ira D. Wallach Division of Art, Prints and Photographs: Picture Collection); 1862.

> Frederick City, September 30th
> It was after nine o'clock last night before we entered the city, and after the longest half hour I have ever experienced, we were put into a hospital, formerly a Presbyterian church. I have learned that the Doctor in charge bears the cognomen "Cornish." He is a gentleman 'of the old school apparently, and I have already experienced kindness from him. He wakened me up last night about twelve o'clock from the soundest sleep I have had since I was shot, and washed and dressed the wound. I showed my appreciation of the act, by resuming my interrupted slumbers, with greater vigor than ever.

> Frederick City, October 4th
> Abraham Lincoln passed through the city today on his return from the review of the army on the upper Potomac. He paid the honor of a visit to a certain General [Union General George L. Hartsuff] who was wounded at Antietam and is stopping opposite here, and to the crowd which collected around the door and cheered him, he made a speech of his usual brevity. He was then assisted into an unpretentious one-horse buggy by a field officer

who accompanied him. They passed the hospital, and I had an excellent view of the features of the President. He looked more worn than when I saw him last, and the heavy load he is obliged to carry, amply accounts for that. My present condition is not overly pleasant, but by far better than is his.[19]

New Protestant Episcopal Church (All Saints' Episcopal)
Old Protestant Episcopal Church (All Saints' Episcopal)

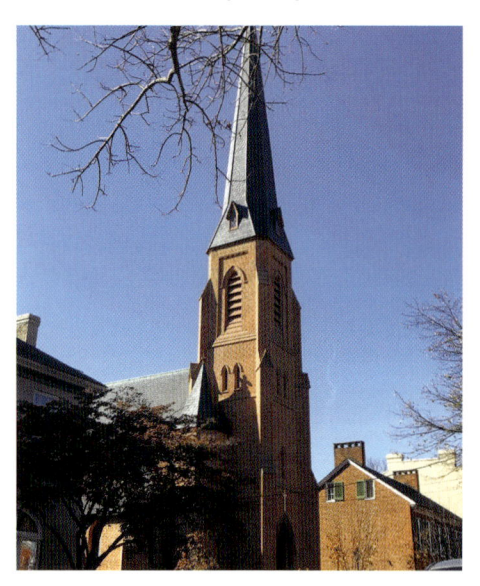

Figure 6.11 **New Church.** The New Protestant Episcopal Church was often referred to as the "New Church" during its use as an Army hospital following the battles of South Mountain and Antietam. It is now the All Saints' Episcopal Church. Courtesy of Scott C. Woodard, US Army Medical Department Center of History and Heritage; November 2017.

Both the "New Church" and the "Old Church," as they were called during their hospital use by the Federal Army, are now combined into the All Saints' Episcopal Church. Like the earlier mentioned German

From the Presbyterian Church, participants can take Record Street (catty-cornered) south for one block to the intersection at Record and Church Streets. After that, they can turn left on Court Street and head east for half a block. The New Protestant Episcopal Church is on the right.

The old courthouse building on the left is now the City Hall of Frederick. It hosted the opening of the Maryland General Assembly on 16 April 1861. Because the building was deemed too small, the rest of the summer sessions took place at Kemp Hall, introduced earlier in the chapter.

Figure 6.12 **New Church Sanctuary.** The New Church maintained some of its structure during its use as a hospital by removing the pews in the sanctuary. Courtesy of Scott C. Woodard, US Army Medical Department Center of History and Heritage; November 2017.

Reformed Church, these two properties belonged to the same congregation. The New Church's building is easily identified with its Gothic Revival style design and characteristic steeple from 1855. Back then, the church chose to remove its pews to establish the hospital inside the sanctuary.

Staff ride participants can continue to travel east and then turn right and go south on Court Street where they will see the Old Church on the right.

Figure 6.13 **Old Church.** The oldest building of the two (compared with the New Church), the Old Protestant Episcopal Church, shortened to Old Church by the patients and staff, stands with a history to tell. It is now the All Saints' Episcopal Church. Courtesy of Scott C. Woodard, US Army Medical Department Center of History and Heritage; November 2017.

Even after the construction of the New Church, the Old Church was used as a parish hall for activities such as Sunday school and church meetings. The Georgian style building, built in 1813, was also used as a hospital facility by the US Army.[20]

Acting Assistant Surgeon Asa A. Bean served at the Old Church on Court Street and wrote home to share his experiences. He described the patients' living spaces on the ground level and in the galleries. While looking out for the safety of one poor soul in the upper levels of the church, Dr Bean directed another delirious typhoid fever patient be removed from the upper levels for fear of his demise from gravity. "I was afraid he would jump out the window or throw himself over the breast works," he said.

The New Church's proximity to the Old Church allowed congregants to visit back and forth. In a letter dated 16 October 1862, Dr Bean wrote,

> My writing is broken in upon by musick in the hospital; I pass out of my room into the gallery & listen to a hymn from 2 first quality singers; then the reading of a chapter in the Bible & a

prayer from the Revd [Reverend] Mr. Diehl & another hymn & they pass on to another hospital & proceed with the same exercises—the inmates of this hospital were much pleased—with 1 or 2 exceptions. . . . It was some while I was upstairs [in the] gallery in the resection of an elbow joint.[21]

Case Studies from *The Medical and Surgical History of the War of the Rebellion (1861-65)*

The following case study documents the care provided to a wounded soldier through his final record of care. The Office of the Surgeon General collected all field medical reports and produced the largest compilation of wartime medical care in its time.

Unable to Shoulder the Burden

CASE: Private G. W. Gentle, Co. E, 5th Ohio, was wounded at Antietam, September 17, 1862, and admitted to hospital No. 3, Frederick, October 1st. Assistant Surgeon J. H. Bill, U.S. A., reported:

"The ball entered to the outside of the right pyramidalis [anterior abdominal] muscle, passed outward and downward in front of the femur, and emerged at a spot in the integument [covering] corresponding to the insertion of the gluteus maximus. Nothing happened in this case, and no injury of grave character was suspected. On the 23d of October, however, a hæmorrhage from both wounds took place. It was dark in color and readily checked by a tampon. Accordingly no action was taken, and, on the 25th, it

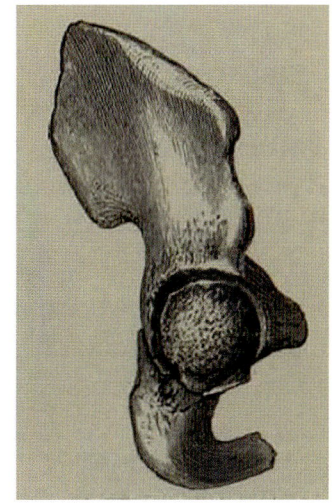

Figure 6.14 **Head of Femur.** The right os innominatum (fused bones of the pelvis) and head of the femur of Private G. W. Gentle. Otis, George A. *The Medical and Surgical History of the War of the Rebellion (1861-65) Part III, Volume II, Surgical History.* Washington, DC: US Government Printing Office; 1883:77.

recurred. It was now found that the thigh and hip were much swollen, an abscess present, seated in the track or the wound, and hæmorrhage evidently due to the ulceration of some vessel. In consultation with Surgeon H. S. Hewit, U. S. V., ligation of the external iliac [lower body regions] was determined on. The man, however, refused the operation, and as the hæmorrhage externally had ceased, it was considered proper to wait. On October 27th, the hæmorrhage returned and the man wanted the artery tied. He was nearly moribund, and the case otherwise being unpromising still, the operation was undertaken by an incision parallel to but outside of the epigastric artery. The external iliac was found and tied without any difficulty. Previous to tying the ligature it was intended to lay open the abscess and search for the bleeding vessel, knowing that the circulation could be controlled.

But at this stage of the operation the shock to the patient was so severe that it was necessary to finish all operative interference as soon as possible. The patient survived the operation only twenty-four hours. The autopsy showed the external iliac tied a quarter of an inch below the internal iliac, and the vein and peritoneum [membrane surrounding abdomen] uninjured. A syringe was introduced into the femoral artery below the origin of the profunda [deep artery in the thigh] and water thrown upward, but the bleeding vessel was not discovered, notwithstanding a careful dissection and prolonged search. The syringe was now introduced into the internal iliac and the water thrown downward, but with no better result. Failing thus to discover an ulcerated vessel on the cadaver, what likelihood would there have been of a successful search for the bleeding point on the living patient? The operation performed was unquestionably the proper one, as it checked the hæmorrhage, and was the only feasible method of doing this."

The specimen represented in [Figure 6.14] was contributed by Assistant Surgeon J. H. Bill, and consists of "the right os innominatum [fused bones of the pelvis] and head of the femur. The joint was opened and the ischium [curved bone of pelvis] at the lower border of the acetabulum [hipbone socket] contused by a musket ball which escaped through the gluteus maximus. The articular surfaces are eroded, but the implication of the joint was not suspected during life."[22]

Case Studies from *The Medical and Surgical History of the War of the Rebellion (1861-65)*

The following case study documents the care provided to a wounded soldier through his final record of care. The Office of the Surgeon General collected all field medical reports and produced the largest compilation of wartime medical care in its time.

Gunshot in the Elbow is not Humorous

CASE: Private G. L. Essick, Co. K, 7th Pennsylvania Reserves, aged 19 years, was wounded at Antietam, September 17, 1862, and entered Hospital No. 3, Frederick, on October 1st, where he underwent excision of the elbow joint. Assistant Surgeon J. H. Bill, U. S. A., who performed the operation, transmitted the following report of the case:

"Gunshot wound of right elbow joint. I saw the case on the 3d of October. The patient was not in a good state of health. The olecranon [bony part of elbow] was destroyed in part, the internal condyle [round end] wounded, and the ulnar [long bone in forearm] nerve divided. The indication here was for constitutional treatment and then a resection. On the 29th of October, the patient's condition being much improved, a resection of the joint was made, a T-shaped incision being employed. The condyles of the humerus and the olecranon process were removed. The insertion of the brachialis anticus [muscle that flexes elbow joint] was preserved. None of the radius was removed. No hæmorrhage occurred. The sensation in the outer fingers, as before the operation, is deficient. Limb placed on a pillow, five wire sutures used, and cold-water

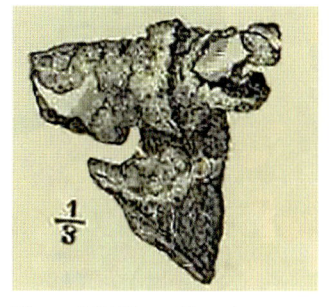

Figure 6.15 **Elbow Drawing.** Bones of the right elbow excised for a shot injury of Private G. L. Essick. Otis, George A. *The Medical and Surgical History of the War of the Rebellion (1861-65) Part II, Volume II, Surgical History*. Washington, DC: US Government Printing Office; 1877:885.

dressings. November 10th, wound united in part; suture removed and limb placed on an angular splint of 90°. Will do well. December 1st, wound nearly healed; sensation in outer fingers greatly improved; patient's health and strength completely restored; limb will be serviceable."

The excised parts, including the coronoid process [triangular eminence from long bone in forearm], were contributed to the Museum [National Museum of Civil War Medicine] by the operator, and are represented in the annexed wood-cut [Figure 6.15]. They are carious [decayed], and the line of section in the ulna [long bone in forearm] is exceedingly oblique. On January 24th, the patient was transferred to Hospital No. 1, whence he was discharged April 4, 1863, and pensioned.

Examiner T. B. Reed, of Philadelphia, September 13, 1865, certified: "Gunshot wound right elbow joint; resection of upper third of ulna. Use and power of arm a good deal impaired." The Philadelphia Examining Board, consisting of Drs. H. E. Goodman, J. Collins, and T. H. Sherwood, reported, May 27, 1874: "Ball struck behind internal condyle of humerus, passing through and out in front. Result, fracture of condyle, with loss of the ends of the bones, producing considerable deformity, leaving the joint partially anchylosed [fused]. Cannot extend beyond angle of forty-five degrees. Rotation seriously impaired. Disability rated three-fourths. Ball re-entered abdomen three inches above crest of right ilium [bone of upper half of pelvis], and out one inch to left of median line, above umbilicus [navel]. Complains of burning sensation in track of wound in bad weather. Does not cause him any serious trouble otherwise. Disability rated one-fourth." The pensioner was paid June 4, 1875.[23]

From the Old Protestant Episcopal Church, staff ride participants can head south to the intersection of Court and Patrick Streets. After that, they can turn left and proceed east for about three blocks to the final stop—the National Museum of Civil War Medicine (Chapter 6, Stop 4)—on the right.

6/4 STOP 4: NATIONAL MUSEUM OF CIVIL WAR MEDICINE

From this point in the study, participants have explored the Battle of Antietam evacuation channels that originated from the point of injury on the field of battle to the general hospital in Frederick, Maryland. The National Museum of Civil War Medicine (**Chapter 6, Stop 4**) provides a closing review of US Civil War medicine and an opportunity to present the staff ride discussion (integration) phase. Participants are encouraged to "travel back in time" along the artifacts and testimony presented here. The stories of individuals make up the lessons for groups of people. Students gain from the conscious decision not to repeat mistakes. Just as important, however, it is also wise counsel to replicate that which worked well.

Visual stories are displayed in words and in scenes that allow viewers to immerse themselves in the action. The case studies throughout the reading, as stated earlier, were documented in medical histories and collected by the Office of the Army Surgeon General. The specimens discussed in the records also point to a final disposition afterward. Bone and tissue samples were collected and forwarded to the US Army Medical Museum to serve as a laboratory for the learning of diseases and wounds. Many of these specimens are now displayed at the National Museum of Civil War Medicine. The US Army Medical Museum was later transformed into the National Museum of Health and Medicine, and the Office of the Army Surgeon General's Library later became the National Library of Medicine.

Confederate Private R. P. Hughes, Company D, 50th Georgia Infantry Regiment, was wounded on

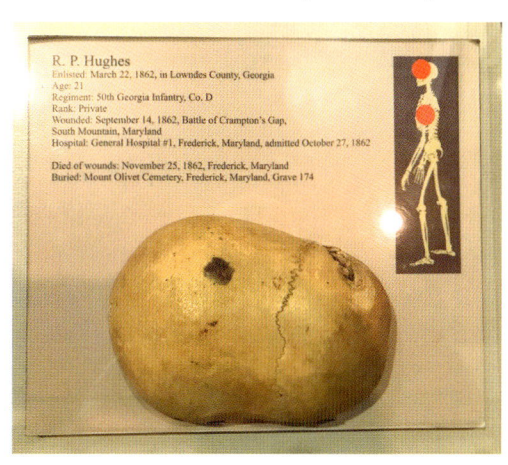

Figure 6.16 **Skull Photo.** Portion of the skull from Confederate Private R. P. Hughes treated at General Hospital Number 1 in Frederick, Maryland. Courtesy of Scott C. Woodard, US Army Medical Department Center of History and Heritage; November 2017.

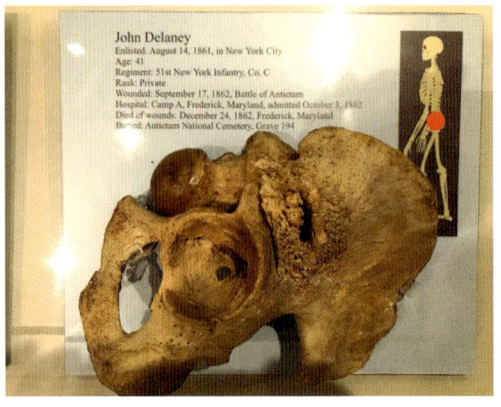

Figure 6.17 **Hip Photo.** Portion of the hip from Federal Private John Delaney treated at Camp A General Hospital in Frederick, Maryland. Courtesy of Scott C. Woodard, US Army Medical Department Center of History and Heritage; November 2017.

14 September 1862 at the Battle of South Mountain. A lead ball struck his forehead and fractured his left frontal bone. Another ball made its mark in his upper arm. After his injuries, Hughes was eventually evacuated to General Hospital Number 1, the old Hessian Barracks in Frederick. The soldier suffered from diarrhea and infection and died 25 November 1862.

Federal Private John Delaney, 51st New York Infantry, took a Minie ball to his hip on 17 September 1862 at the Battle of Antietam. He was initially treated in a field hospital closer to the battlefield, but was eventually transferred to the Frederick Tent Hospital, Camp A. Although the wound produced a great deal of pus, Delaney persevered and began walking again. However, an unfortunate fall caused his health to deteriorate and he died on Christmas Eve 1862.

Reflection

At this stop, at the National Museum of Civil War Medicine, staff ride participants will be given the opportunity to reflect upon the vicarious patient and provider experiences seen throughout this reading. Here are some questions to consider:

- What kind of thoughts do the sights and sounds of the battlefield conjure?
- How does terrain affect how commanders make decisions?
- What principles of battlefield medicine practiced in 1862 are seen today?
- What areas in military medicine needed improvement following the Battle of Antietam?
- How did the case studies compare to your previous ideas of Civil War medicine?

- How can military medicine help combatant commanders in carrying out the overall mission?
- In what way are patients, medical staff members, and combatant commanders considered to be "customers"?
- What qualities make up a good staff officer?
- What is medical readiness and how did Major (Dr) Jonathan Letterman perform?

Introduced earlier in Chapter 4 at Keedysville, Maryland, Dr Thomas T. Ellis wrote about the battle studied in these pages soon after it concluded.

> The total loss, in both battles [South Mountain and Antietam], is fourteen thousand seven hundred and ninety-four. The immense number of wounded to be cared for—over ten thousand—nearly one half of whom required surgical operations, and all requiring care and hospital accommodation, will give the reader some idea of the amount of duty and responsibility devolving on the medical department of the army, and will show the necessity of placing at its head a man of unquestioned ability, and incorruptible integrity, who cannot be approached by parties interested in obtaining contracts for supplying the hospital stores, medicines, instruments, and appliances, the cost of which, for the last year, has exceeded twelve millions of dollars.[24]

Dr Ellis' assessment and thoughts point toward the man of Jonathan Letterman. Even though he states the number of soldiers lost in battle is lower than modern estimates, his point is still made. Dr Letterman executed the mission by ensuring the readiness of his medical department. By this measure, he saved lives and made history.

Another physician from the 20th century, Army Major General (Dr) Paul R. Hawley, weighed in on the legacy of the former medical director.

> I was the Chief Surgeon of the European Theater of Operations during World War II, a position similar to that of Letterman in the Army of the Potomac. At that time I often wondered whether, had I been confronted with the primitive system which Letterman fell heir to at the beginning of the Civil War, I could have developed as good an organization as he did. I doubt it. There was not a day during World War II that I did not thank God for Jonathan Letterman.[25]

References

1. Woodward JJ. *The Medical and Surgical History of the War of the Rebellion (1861-65), Part I, Volume I, Medical History, Appendix to Part I, Containing Reports of Medical Directors, and Other Documents*. Washington, DC: US Government Printing Office; 1870:98.
2. Letterman J. *Medical Recollections of the Army of the Potomac*. New York, NY: D. Appleton & Co; 1866:44–45.
3. Woodward JJ. *The Medical and Surgical History of the War of the Rebellion (1861-65), Part I, Volume I, Medical History, Appendix to Part I, Containing Reports of Medical Directors, and Other Documents*. Washington, DC: US Government Printing Office; 1870:105.
4. Woodward JJ. *The Medical and Surgical History of the War of the Rebellion (1861-65), Part I, Volume I, Medical History, Appendix to Part I, Containing Reports of Medical Directors, and Other Documents*. Washington, DC: US Government Printing Office; 1870:107.
5. Reimer T. *One Vast Hospital: The Civil War Hospital Sites of Frederick, Maryland, and after Antietam*. Frederick, MD: The National Museum of Civil War Medicine; 2001:72–74.
6. Diary of Cyrus Bacon. Accessed June 15, 2020(23). https://www.civilwardigital.com/
7. Diary of Cyrus Bacon. Accessed June 15, 2020(24,28). https://www.civilwardigital.com/
8. Otis GA. *The Medical and Surgical History of the War of the Rebellion (1861-65), Part III, Volume II, Surgical History*. Washington, DC: US Government Printing Office; 1883:537–538.
9. Otis GA. *The Medical and Surgical History of the War of the Rebellion (1861-65), Part III, Volume II, Surgical History*. Washington, DC: US Government Printing Office; 1883:587.
10. Reimer T. *One Vast Hospital: The Civil War Hospital Sites of Frederick, Maryland, and after Antietam*. Frederick, MD: The National Museum of Civil War Medicine; 2001:62.
11. *275th Anniversary Commemorative Booklet: Celebrating a Milestone - Growing in Grace for 275 Years*. Frederick, MD: Evangelical Lutheran Church;2013.
12. The Staff and Volunteers of the Hood College Archives. *Hood College*. Charleston, SC: Arcadia Publishing;2013:9–13.
13. Oates SB. *A Woman of Valor: Clara Barton and the Civil War*. New York, NY: The Free Press; 1994:97–98.
14. Otis GA. *The Medical and Surgical History of the War of the Rebellion (1861-65), Part II, Volume II, Surgical History*. Washington, DC: US Government Printing Office; 1877:46–47.
15. Maryland State Archives. *The General Assembly Moves to Frederick; 1861*. Accessed June 15, 2020. https://msa.maryland.gov/msa/stagser/s1259/121/7590/html/0000.html
16. Douglas HK. *I Rode with Stonewall*. St. Simons Island, GA: Mockingbird Books; November 1987:150–151.
17. Reimer T. *One Vast Hospital: The Civil War Hospital Sites of Frederick, Maryland, and after Antietam*. Frederick, MD: The National Museum of Civil War Medicine; 2001:47.
18. Reimer T. *One Vast Hospital: The Civil War Hospital Sites of Frederick, Maryland, and after Antietam*. Frederick, MD: The National Museum of Civil War Medicine; 2001:59.
19. Johnson CF. *The Long Roll*. East Aurora, NY: The Roycrofters;1911:200–203.

20. Reimer T. *One Vast Hospital: The Civil War Hospital Sites of Frederick, Maryland, and after Antietam*. Frederick, MD: The National Museum of Civil War Medicine; 2001:55–56.
21. Wallace DH. Two Centuries of Service: All Saints' Church on Court Street. Frederick, MD: All Saints' Church; 2014:17.
22. Otis GA. *The Medical and Surgical History of the War of the Rebellion (1861-65), Part III, Volume II, Surgical History*. Washington, DC: US Government Printing Office; 1883:77.
23. Otis GA. *The Medical and Surgical History of the War of the Rebellion (1861-65), Part II, Volume II, Surgical History*. Washington, DC: US Government Printing Office; 1877:885.
24. Ellis TT. *Leaves from the Diary of an Army Surgeon; or Incidents of Field Camp, and Hospital Life*. New York, NY: John Bradburn, publisher; 1863:305.
25. Bollet AJ. *Civil War Medicine: Challenges and Triumphs*. Tucson, AZ: Galen Press; 2002:97.

SELECTED ANTIETAM MEDICAL HISTORIES

> " To care for him who shall have borne the battle. . . ."
>
> —President Abraham Lincoln
> Second Inaugural Address, 4 March 1865

Two Inches Too Short

Case Studies from *The Medical and Surgical History of the War of the Rebellion (1861-65)*

The following case study documents the care provided to a wounded soldier through his final record of care. The Office of the Surgeon General collected all field medical reports and produced the largest compilation of wartime medical care in its time.

CASE: Private O. L. Bell, Co. D, 1st Delaware, aged 19 years, was wounded in the right leg, at Antietam, September 17, 1862, and admitted to hospital at Frederick [Hospital #1[1]] ten days afterwards.

Acting Assistant Surgeon W. S. Adams forwarded the following history: "An examination revealed extensive comminuted fracture of both bones of right leg at the upper portion of the lower third, a transverse fracture at the upper third, and an oblique fracture running down to within two inches of the external malleolus [bony projection at the side of the ankle]. The limb was in Smith's anterior splint [a type of splint developed by Professor Nathan R. Smith], which had been badly applied; and a piece of adhesive plaster, which had been placed just above the knee, had been allowed to receive the weight of the limb for six weeks. The result was that it cut through the skin, fascis [connective tissue], and to a considerable extent into the muscles, the incision made being seven inches long and at its middle two and a half inches broad.

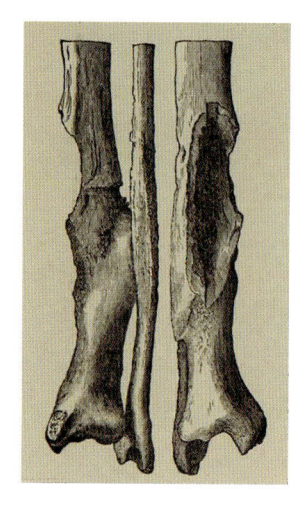

Figure 1 **Bones of Right Leg.** A posterior and anterior view of the bones of the right leg. Otis, George A. *The Medical and Surgical History of the War of the Rebellion Part III, Volume II, Surgical History.* Washington, DC: US Government Printing Office; 1883:544.

After taking it off the limb was readjusted in the same splint, in which it remained about three weeks longer, when it was placed in a fracture box. After union the leg showed two inches shortening. The patient's condition had remained good throughout, but on the morning of December 26th he had a severe chill, followed at 10 A.M. by considerable fever. On the next day there was some evidence of erysipelas [bacterial infection, St. Anthony's fire] on the leg and thigh, and three grains of quinine were prescribed every three hours, also fifteen drops of tincture of chloride of iron [tonic to raise pulse and increase secretions] every four hours. On the following day erysipelas was very evident and extended from the ankle to the hip; limb hot and much swollen; pulse 130; tongue furred and bowels constipated. Saline cathartics [purgatives] were now ordered and lead and opium wash was applied. On December 29th the patient was no better and there was total loss of appetite. The limb was now suspended in Smith's anterior splint so as to allow a free passage of air beneath and to facilitate the application of local remedies to all parts affected. After this the patient did quite well until January 10, 1863, when there was some evidence of an abscess on the anterior part of the thigh, but no distinct fluctuation could be recognized; patient having no pain and feeling quite well, his appetite having returned for some days. The quinine was now stopped and brandy and tonics were continued. On January 14th an extensive abscess was opened on the anterior part of the thigh and about a quart of pus was evacuated, after which the cavity was syringed with tepid water and a bandage was applied to the entire limb. One week later a solution of zinc was ordered to be used for syringing. Subsequently the patient continued to do well, requiring no treatment, and on February 10, 1863, he was discharged, the walls of the abscess having become adherent and the ulcer nearly cicatrized [scarred]."

The man subsequently re-enlisted in the 1st Delaware Cavalry and served for fifteen months, when the wound re-opened. He then passed through various hospitals, being ultimately discharged for disability, from Jarvis Hospital, Baltimore, June 15, 1865, and pensioned.

Examiner I. Jump, of Dover, Delaware, certified July 1, 1871: "There is a large open sore some four or five inches long and the skin or flesh on most of the leg is very much discolored, the discharge being very offensive except when counteracted by disinfectants. The pensioner had to take to his bed last February, being unable to bear any weight on the limb and suffering very much with it. His physician, who had served in the army, insisted on taking the leg off. I am of the opinion that it never will be healed; but l have advised that the diseased portions of the bones be removed; it is barely possible he may recover."

Dr. J. F. M. Forwood, of Chester, Pennsylvania, who subsequently, on August 19, 1872, amputated the leg seven inches below the knee joint, communicated, in connection with the case, that after receiving his final discharge from service the man "roamed about and had one or two operations performed for his relief, staying some fifty days in St. Joseph's Hospital, Philadelphia, and finally drifting here, where I amputated his limb."

At an examination of the stump in August, 1879, Examiner Jump reported: "There has been ulceration for ten months continuously, sometimes confining him to his bed." The pensioner was paid March 4, 1881. The amputated bones of the leg [Figure 1] [Bone and tissue samples were collected and forwarded to the US Army Medical Museum to serve as a laboratory for the learning of disease and wounds.], together with a photograph of the pensioner, represented in the wood-cuts [Figure 2] were contributed to the Museum by the operator.[2]

Figure 2 **Stump of Amputee.** An image of Private Bell after his amputation was submitted to the US Army Medical Museum. Otis, George A. *The Medical and Surgical History of the War of the Rebellion (1861-65), Part III, Volume II, Surgical History.* Washington, DC: US Government Printing Office; 1883:544.

'I am Still in the Land of the Living'

Case Studies from *The Medical and Surgical History of the War of the Rebellion (1861-65)*

The following case study documents the care provided to a wounded soldier through his final record of care. The Office of the Surgeon General collected all field medical reports and produced the largest compilation of wartime medical care in its time.

The following case history, introduced in Chapter 1, provides an example of how Private Edson D. Bemis' further injuries and recovery were recorded.

CASE: Private Edson D. Bemis, Co. K, 12th Massachusetts Volunteers, was wounded in Antietam by a musket ball which fractured the shaft of his left humerus. The fracture united kindly, with very slight angular displacement and quarter of an inch shortening. Promoted to be corporal, Bemis received May 6th, 1864, at the battle of the Wilderness a wound from a musket ball in the right iliac fossa [part of the hip bone]. He was treated in the Chester Hospital, near Philadelphia. There was extensive sloughing about the wound, but it ultimately healed entirely, leaving a large cicatrix [scar], parallel with Poupart's ligament [fibrous band from pelvis to lower limb].

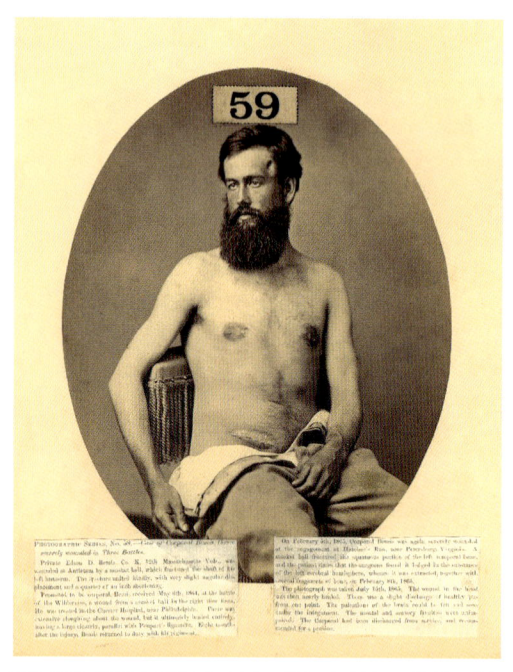

Figure 3 **Private Bemis.** Union Private Edson D. Bemis' original photograph as documented and recorded at the US Army Medical Museum in Washington, DC. Source: US National Library of Medicine.

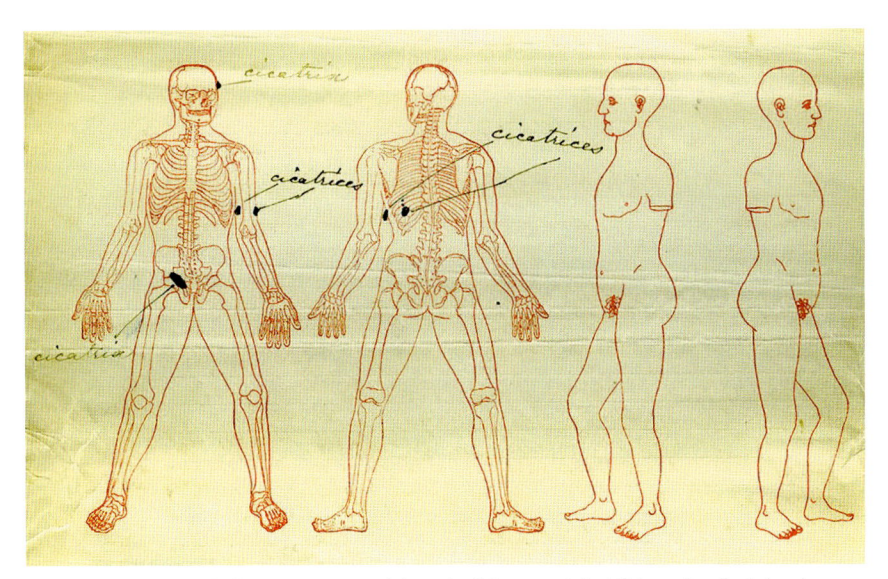

Figure 4 **Anatomical Diagram.** An examining physician annotated this anatomical drawing to show Union Private Edson D. Bemis' injuries. This drawing was part of his Application for Increase of Pension from 10 March 1886. Source: Sharp, Rebecca K. and Wing, Nancy L. I Am Still in the Land of the Living: The Medical Case of Civil War Veteran Edson D. Bemis. *Prologue Magazine*. Washington, DC: National Archives and Records Administration;Spring 2011.

Eight months after the injury, Bemis returned to duty with his regiment. On February 5th, 1865, Corporal Bemis was again severely wounded at the engagement at Hatcher's Run, near Petersburg, Virginia. Surgeon A. Vanderveer, 66th New York Volunteers, reports that the ball entered a little outside of the left frontal protuberance [bump], and passing backward and upward, removed a piece of the squamous portion of the temporal bone [outside of the cranium], with brain substance and membranes. When the patient entered the hospital of the 1st division of the Second Corps, brain matter was oozing from the wound. There was considerable hæmorrhage, but not from any important vessel. Respiration was slow; the pulse 40; the right side was paralyzed and there was total insensibility. On February 8th, the missile was removed from the substance of the left hemisphere, by Surgeon Vanderveer. It was a conoidal musket ball, badly battered. The patient's condition at once improved. He told the surgeon his name, and seemed conscious of all that was going on about him. Water dressings were

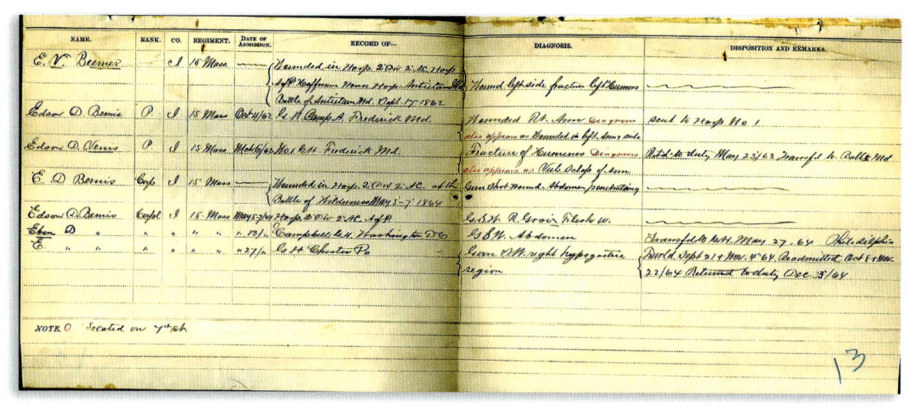

Figure 5 **Pension File.** Union Private Edson D. Bemis' pension file recorded his wounds, diagnoses, and his physician's comments. Source: Sharp, Rebecca K. and Wing, Nancy L. I Am Still in the Land of the Living: The Medical Case of Civil War Veteran Edson D. Bemis. *Prologue Magazine*. Washington, DC: National Archives and Records Administration;Spring 2011.

applied, and an ingeniously arranged sponge absorbed the discharge from the wound. He was kept on a very light diet and remained very quiet for ten days, answering direct questions, but indisposed to continue a conversation. He had no convulsions and his sleep was not disturbed by delirium. About February 18th, a marked improvement was manifest. The patient conversed freely, and the wound was rapidly cicatrizing [healing with scar], and the hemiplegia [paralysis on one side] had entirely disappeared. On February 28th he was able to walk about the ward. On March 18th the wound was nearly healed. The patient was sent northward on a hospital transport to Fort Richmond, New York Harbor. He recovered perfectly, and in May was furloughed, and on May 18th he wrote to Dr. Vanderveer, that he was doing well at his home in Huntington, Massachusetts, suffering only slight dizziness in going out in the hot sun. In July he went to Washington to apply for a pension, and entered Campbell Hospital. He was discharged on July 13th, 1865, on surgeon's certificate of disability. At this date he was photographed at the Army Medical Museum. The wound in the head was then nearly healed. There was a slight discharge of healthy pus from one point. The pulsations from the brain could be felt though the integument [outer layer]. The mental and sensory faculties were unimpaired. The corporal had been discharged from service and recommended for a pension. The plate opposite is a very accurate copy of the photograph (Figure 3), which is numbered 58 [published as 59] of the surgical series, A. M. M. [Army Medical Museum]. Mr. Bemis was pensioned at eight dollars per month.

On October 30th, 1870, he wrote to the editor of the surgical history from his home in Suffield, Connecticut, as follows: "I am still in the land of the living. My health is very good considering what I have passed through at Hatcher's Run. My head aches some of the time. * * I am married and have one child, a little girl born last Christmas. My memory is affected, and I cannot hear as well as I could before I was wounded."[3]

REFERENCES: SELECTED ANTIETAM MEDICAL HISTORIES

1. Reimer T. *One Vast Hospital: The Civil War Hospital Sites of Frederick, Maryland, and after Antietam.* Frederick, MD: The National Museum of Civil War Medicine; 2001: Appendix A.
2. Otis GA. *The Medical and Surgical History of the War of the Rebellion (1861-65), Part III, Volume II, Surgical History.* Washington, DC: US Government Printing Office; 1883:544–545.
3. Otis GA. *Medical and Surgical History of the War of the Rebellion (1861-65), Part I, Volume II, Surgical History.* Washington, DC: US Government Printing Office; 1870:162.

ANTIETAM CASUALTIES

Casualty Does Not Equal Deceased

Casualties include three categories: 1) deceased; 2) wounded; and 3) missing or captured. In general terms, casualties of Civil War battles included 20% dead and 80% wounded. Of the soldiers who were wounded, about one out of seven died from their wounds. Over two-thirds of the approximately 622,000 men who gave their lives in the Civil War died from disease, not from battle.

Table 1 **Approximate Antietam Casualties by Type**

STATUS	UNION	CONFEDERATE	TOTAL
Killed	2,100	1,550	3,650
Wounded	9,550	7,750	17,300
Missing/Captured	750	1,020	1,770
TOTAL	12,400	10,320	22,720

Note: Because of the catastrophic nature of the Battle of Antietam, exact numbers of casualties were virtually impossible to compile. The sources for these figures are The Official Records of the War of the Rebellion and the Antietam Battlefield Board. Formed in 1894, the Antietam Battlefield Board mapped the battlefield and established the movement of troops with markers on the field.

Table 2 **Approximate Casualties by Phase of the Battle of Antietam**

		UNION	CONFEDERATE	TOTAL
Morning Phase	Engaged	23,600	20,100	43,700
	Casualties	7,280	6,580	13,860
Midday Phase	Engaged	10,000	6,800	16,800
	Casualties	2,900	2,600	5,500
Afternoon Phase	Engaged	13,800	7,150	20,950
	Casualties	2,600	1,120	3,720

Worst Civil War Battles

Antietam was the bloodiest one-day battle of the Civil War. But there were other battles, lasting more than one day, in which more men fell. The numbers below are the total casualties for both sides.

Source: https://www.nps.gov/anti/learn/historyculture/casualties.htm

Table 3 **Approximate Number of Casualties in the Worst Civil War Battles**

BATTLE	NUMBER OF CASUALTIES	DATE
Gettysburg	51,000	1–3 July 1863
Chickamauga	34,624	18–20 Sep 1863
Wilderness	29,800	5–7 May 1864
Chancellorsville	24,000	1–4 May 1863
Shiloh	23,746	6–7 Apr 1862
Stones River	23,515	31 Dec 1862 to 1 Jan 1863
Antietam	22,717	17 Sep 1862
2nd Manassas	22,180	28 Aug to 1 Sep 1862

Source: The American Battlefield Protection Program.
Source: https://www.nps.gov/anti/learn/historyculture/casualties.htm

MEDICAL OFFICERS OF THE ARMY OF THE POTOMAC AT ANTIETAM

Medical Director	Surgeon Jonathan Letterman, USA
Medical Inspector	Lieutenant Colonel Edward P. Vollum, USA
Medical Purveyor	Assistant Surgeon Thomas J. McMillin, USA
Medical Director, First Corps	Surgeon John Theodore Heard, USV
First Division	Surgeon Peter Pineo, USV
Second Division	Surgeon Nathaniel Richard Moseley, USV
Third Division	Surgeon William King, USA
Medical Director, Second Corps	Surgeon Alexander Nelson Dougherty, USV
First Division	Surgeon John Howard Taylor, USV
Second Division	Surgeon Houston, USV
Third Division	Surgeon Gabriel Grant, USV
Medical Director, Fifth Corps	Surgeon John McNulty, USV
First Division	unknown
Second Division	unknown
Medical Director, Sixth Corps	Surgeon William J. H. White, USV (killed)
	Surgeon Charles O'Leary, USV
First Division	unknown
Second Division	unknown
Medical Director, Ninth Corps	Surgeon William Henry Church, USV
First Division	unknown
Second Division	Surgeon Alexander T. Watson, USV
Third Division	Surgeon Henry Wheaton Rivers, USV
Fourth Division	Surgeon William W. Holmes, USV
Medical Director, Twelfth Corps	Surgeon Thomas Antisell, USA
First Division	Surgeon Artemus Chapel, USV
Second Division	Surgeon Alfred Ball, USV

United States Army (USA)—Regular Army

United States Volunteers (USV)—Volunteers augmenting the Regular Army and state militias

Sources: Duncan, Louis C. *The Medical Department of the United States Army in the Civil War.* Carlisle Barracks, PA: Medical Field Service School; 1931:Chapter V,45.

National Museum of Civil War Medicine in Frederick, Maryland.

INDEX

A

Abbott, Captain Robert A., 72–73
Adams, Dr. W.S., 193–194
Adams, Scout, 5
Agnew, Dr Cornelius R., 137–138
Alcohol abuse, 73
All Saints' Episcopal Church, 180–183
Allen, Private George A., 140
Ambulance Corps
    battle drills, 14
    design of ambulance wagons, 10
    establishment of, 11–13
    implementation of Letterman's plan for evacuation and treatment of wounded, 74
    plan to reorganize ambulance service, 8–9
    regulations for, 11–13
Ambulances
    design of ambulance wagons, 10
    Howard Ambulance, 123–124
    planning of ambulance system, 159
    Rosecrans ambulance wagon, 126–127
    Wheeling ambulance wagon, 126–127
Amputations, 139–142, 163–165, 193–195
Antietam. *See* Battle of Antietam
"Antietam" documentary, xxi
Antietam National Battlefield
    view from Hagerstown Road, 28
    Visitor Center, xx–xxii
"Antietam Visit" movie, xxi
Antisell, Dr Thomas, 202
Ard, Private George Washington Lafayette, 105–108
Army of the Potomac
    headquarters for, 117–119
    Letterman as Medical Director, 8–9, 157–159
    medical officers at Antietam, 202
    medical supplies authorization for, 86–87
Arnold, Private Jacob, 78–79
Artificial respiration technique, 123–124
Austro-Prussian War, 84

B

Bacon, Dr Cyrus, Jr., 162
Ball, Dr Alfred, 202
Barber, Captain Frederick M., 109
Barnard, Lieutenant Harry, 60, 172
Barton, Clara
    arrival at Poffenberger House, 55
    battlefield experiences, 56–60
    photograph, 56
    visit to Frederick hospital, 172
Battle of Antietam. *See also* Field hospitals
    Antietam National Battlefield Visitor Center, xx–xxii
    casualties, 200
    Dunker Church, 36–43
    map, vi, 20
    medical annex, 19
    medical officers of the Army of the Potomac, 202
    Miller Farm Cornfield, 20, 24, 26–35
    North Woods, 21–26
    Sunken Road, 60–66, 131
    view from Hagerstown Road, 28
Battle of South Mountain, 85, 153, 154
Battlefield medicine
    assistance from nongovernmental agencies, 48
    implementation of Letterman's plan for evacuation and treatment of wounded, 74
    Letterman's medical readiness reforms, 4–7, 14–15, 86–87
    logistical support errors, 45
    McClellan's support of medical readiness reform, 5–6
    medical structure, 1
    medical supplies authorization for the Army of the Potomac, 86–87
    medical supply challenges, 46–47
    progress in preparedness, xvii–xviii

Bean, Dr Asa A., 182–183
Bean, Second Lieutenant Charles W., 34
Bell, Private O.L., 193–195
Bemis, Corporal Edson D., 24, 196–199
Bill, Dr J.H., 184, 185
Blake, Private C.C., 43
Bloody Lane, 62–66, 72–73
Bone fractures, 193–194, 196
Bonsall's Academy, 177
Bowes, Private George H., 172–174
Brain injuries, 197–199
Bronson, Dr George, 97
Burnside, Major General Ambrose E., 95, 104
Burnside Bridge
 burial sites, 103
 General Burnside's assault to take bridge, 104–108
 letter home, 97
 renaming of, 95
 view of, 96, 104
 wounds and injuries, 98–109

C

Camp A General Hospital, 188
Campbell, Dr Charles Fitz Henry, 138
Care packages, 48–49
Casualties. *See* Wounds and injuries
Chapel, Dr Artemus, 202
Cherbonnier, Dr A.V., 163, 166
Child, Dr William, 33, 34–35
Christ Reformed United Church of Christ Hospital, 154–155
Church, Dr William Henry, 202
Civil War. *See also* Battle of Antietam; Battlefield medicine
 number of casualties in worst battles, 201
Civil War Campaign Medals, 70
Civil War Medals, 70
Clarkson, Private Benjamin, 82
Coffin, Charles Carleton, 118–119
Collins, Dr J., 186
Coppersmith Hall, 176–177

Cornfield, D.R. Miller Farm
 battle, 24
 Freemasons and, 32–34
 Irish Brigade, 30
 letter home, 35
 map, 20, 26
 modern view of, 27
 wounds and injuries, 28–32
Corps Hospitals, 149–156
Cost, Samuel, 139
Cross, Colonel E.E., 33
Curran, Dr Richard J., 67–70
Custer, Captain George Armstrong, 5

D

Delaney, Private John, 188
Derby, Captain Richard, 49
Diarrhea, 37
Diet issues, 37
Dimon, Dr Theodore, 99–102, 108
Dougherty, Dr Alexander Nelson, 202
Douglas, Henry Kid, 175–176
Douglas, Reverend Robert, 143
DuBois, Dr H.A., 173
Duffey, Gunner Ed, 31
Dunker Church
 Confederate line in the woods, 40
 food shortage, 37
 Reel Farm field hospital, 42
 as refuge for care and evacuation of wounded soldiers, 36–37
 wounds and injuries, 38–43
Dunn, Dr James L., 55, 59
Dyer, Dr J. Franklin, 119–121

E

East Woods, 23
Edmonds, Sarah Emma, 144
Edon, First Lieutenant, 33
8th Illinois Cavalry, 172–173
11th Connecticut Infantry, 95, 97–98
Ellis, Dr Thomas T., 133, 145, 189

Essick, Private G.L., 185–186
Evacuation Hospitals, 149–156
Evangelical Lutheran Church, 168–170
Evangelical Reformed Church, 175

F

Farley, Dr James L., 123
Ferrero, Colonel Edward, 104
Field hospitals
   Christ Reformed United Church of Christ Hospital, 154–155
   German Reformed Church Hospital, 139–145, 154–155
   implementation of Letterman's plan for evacuation and treatment of wounded, 74
   instructions for organization and staffing of, 88–91
   Keedysville Hospitals, 131–139
   locations, xiv, 53
   Miller Farm Cornfield, 34
   Mount Vernon Reformed Church Hospital, 139–145
   North Woods, 22–23
   Otho J. Smith Farm Hospital, 129–131
   Otto Farm Hospital, 110–111
   Philip Pry House Field Hospital, 117–127
   Poffenberger Farm, 53
   Reel Farm, 42–43
   Roulette Barn Hospital, 76–78, 99–102, 131
   Samuel Pry Mill Hospital, 127–128
   Stone School House, 143
   Straw-Stack Hospital, 67–74
   Zion Lutheran Church Hospital, 153
Fields, Private B., 163–165
15th Massachusetts Infantry, 49
5th New Hampshire Infantry, 32–34
51st New York Infantry, 104
51st Pennsylvania Infantry, 104
1st Brigade, 1st Division, I Corps, 42–43
1st Texas Infantry, 31

Flint, Alvin, Sr., 96
Flint, George, 96
Flint, Private Alvin, Jr., 95–96
Food shortages, 37, 58, 114
Forwood, Dr J.F.M., 195
4th US Artillery, 31
49th Pennsylvania Infantry, 31
Fractures, bone, 193–194, 196
Franco-Prussian War, 84
Frederick, Maryland, 156–160, 168–186
Frederick Academy, 177
Frederick Female Seminary, 170–172
Frederick Presbyterian Church, 177–180
Frederick Tent Hospital, Camp A, 188
French, Brigadier General William H., 37, 62, 131
French's Division. *See* Otho J. Smith Farm Hospital
Fritchie, Barbara, 168

G

Galloway, Mary, 60, 172
Garland, Captain J.M., 151
General hospitals
   Camp A, 188
   establishment of, 157–159
   naming of, 158
   Number 1, 187–188, 193–194
   Number 3, 176–186
   Number 4, 168–176
   Number 5, 161–167
Gentle, Private G.W., 183–184
German Reformed Church Hospital, 139–145, 154–155
Goodman, Dr H.E., 186
Gordon, Colonel John B., 64–65
Grant, Dr Gabriel, 202
Graves, First Lieutenant Janvrin W., 34
Gray, Dr Charles Carroll, 138–139
Grigsby, Colonel Andrew J., 30
Gross, Dr S.W., 43, 98

H

Hagerstown Pike, 25, 38
Hagerstown Road, 28
Hammond, Brigadier General (Dr) William A., 2–3, 5–6
Hard, Dr C., 173
Harris, Dr Elisha, 142–143
Harris, Mrs. John, 71–72
Hartsuff, General George L., 179
Harwood, Dr Frank, 60
Hawkins' Zouaves, 106, 107, 112–114, 136, 178
Hawley, Major General (Dr) Paul R., 190
Head wounds, 42–43
Headquarters for the Army of the Potomac, 117–119
Heard, Dr John Theodore, 202
Herzog, Sergeant Joseph, 31
Hewit, Dr Henry S., 161, 163–164, 184
Hill, Major General Daniel H., 62, 64
Hinks, Colonel Edward, 120
Hoffman Hospital, 120
Holmes, Dr William W., 202
Holstein, Anna, 48
Holt, Surgeon Daniel M., 28–29
Hood, Lieutenant General John B., 21, 37
Hooker, Major General Joseph, 21, 24, 55, 121–122
Hoover, Dr George W., 73
Hospitals
  All Saints' Episcopal Church, 180–183
  Christ Reformed United Church of Christ Hospital, 154–155
  D.R. Miller Farm Hospital, 34
  Evangelical Lutheran Church, 168–170
  Frederick Female Seminary, 170–172
  Frederick Presbyterian Church, 177–180
  general hospitals, 157–188
  German Reformed Church Hospital, 139–145, 154–155
  Hoffman Hospital, 120
  Keedysville Hospitals, 131–139
  locations near Antietam National Battlefield, xxii
  map of locations, xiv, 53
  Mount Vernon Reformed Church Hospital, 139–145
  New Protestant Episcopal Church, 180–183
  North Woods, 22–23
  Old Protestant Episcopal Church, 180–183
  Otho J. Smith Farm Hospital, 129–131
  Otto Farm Hospital, 110–111
  Philip Pry House Field Hospital, 117–127
  Poffenberger Farm, 53
  Presbyterian Church, 177–180
  Reel Farm, 42–43
  Roulette Barn Hospital, 76–78, 99–102, 131
  Samuel Pry Mill Hospital, 127–128
  Sherrick Farm Hospital, 110–111
  Smoketown Hospital, 50–51
  Stone School House, 143
  Straw-Stack Hospital, 67–74
  US Hotel Hospital, 160
  Winchester Seminary, 170–172
  Zion Lutheran Church Hospital, 153
Houston, Dr, 137, 202
Howard, Dr Benjamin Douglas, 121–124
Howard Ambulance, 123–124
Hughes, Private R.P., 187–188
Humphreys, Dr George, 106, 107, 113
Humphreys, General Andrew A., 5
Hunt, General Henry J., 5
Hurd, Dr Anson, 130

I

Injuries. *See* Wounds and injuries
Irish Brigade, 30, 63, 65, 71–72
Irwin, Major (Dr) Bernard, 84
IX Corps, 104–105, 108

J

Jackson, Major General Thomas "Stonewall" J., 21, 37, 118, 175–176
Jarvis Hospital, Baltimore, Maryland, 173
Jesuit Seminary, 161
Johnson, Clifton, 110
Johnson, Private Charles F., 101, 112–114, 178–180
Jump, Dr I., 195

K

Keedysville, Maryland, 117–127, 132
Keedysville Hospitals, 131–145
Kemp Hall, 175, 180
King, Dr William, 202
King, Private Charles, 31
Kirby, Lieutenant, 135

L

Ladies' Aid Society of Philadelphia, 71–72
Lee, General Robert E.
   health of, 16
   letter of appreciation to troops under his command, 93
   riding along Confederate lines, 64
Letterman, Major (Dr) Jonathan
   comments on medical logistical support errors, 45
   comments on medical supply challenges, 46–47, 135–136
   development of the Rosecrans ambulance wagon, 126
   establishment of general hospitals, 157–159
   Headquarters for the Army of the Potomac, 118
   inspecting potential hospital locations, 129, 131
   the Letterman Plan, 84–85
   as Medical Director of the Army of the Potomac, 8–9, 157–159, 202
   medical readiness reform, 4–7, 14–15, 86–87, 189–190
   medical supplies authorization for the Army of the Potomac, 86–87
   organization and staffing of field hospitals, 88–91
   plan for evacuation and treatment of wounded, 74, 149–152
   support of Ambulance Corps establishment, 8–10
The Letterman Plan, 84–85
Letters home
   Bronson, Dr George, 97
   Child, Dr William, 35
   Derby, Captain Richard, 49

Lincoln, Abraham
   visit to Frederick, Maryland, 179–180
   visit with soldiers near the Antietam battlefield, 5
Little Antietam Creek, 133
Locke, Colonel Frederick T., 5
Longstreet, Major General James, 66
Lower Bridge. *See* Burnside Bridge
Ludlam, Captain J.D., 172
Lutheran Church, 168–170

M

MacKenzie, T.G., 173
Mansfield, General Joseph K., 21
Maps
   Battle of Antietam, vi, 20
   Frederick, Maryland, 156
   hospital locations, xiv, 53
   Keedysville, Maryland, 132
   medical annex to the Battle of Antietam, 19
   Middletown, Maryland, 148
   Miller Farm Cornfield, 20, 24, 26
   North Woods, 26
   Roulette Barn Hospital, 76–78
McClellan, Major General George B.
   establishment of Ambulance Corps, 10–13
   Headquarters for the Army of the Potomac, 117–119
   health of, 17
   letter of appreciation to troops under his command, 92
   support of medical readiness reform, 5–6
McLocklin, Paul J., 107, 108
McMillin, Dr Thomas J., 202
McNulty, Dr John, 202
Meagher, Brigadier General Thomas Francis, 71–72, 73
Medals of Honor, 70
Medical officers, 202
Medical readiness. *See* Battlefield medicine
Messaging system, 155
Methodist Protestant Church, 176–177
Middletown, Maryland, 148–155
Milhau, Dr J.J., 159
Miller, Captain M.B., 66

Miller, D.R., 20, 24, 27–35
Miller, Private George D., 25
Miller Farm Cornfield
    battle, 24
    D.R. Miller Farm Hospital, 34
    Freemasons and, 32–34
    Irish Brigade, 30
    letter home, 35
    map, 20, 26
    modern view of, 27
    wounds and injuries, 28–32
Miller Farm Hospital, 34, 178
Monteith, Captain George, 5
Morell, General George W., 5
Mosby, John, 39
Moseley, Dr Nathaniel Richard, 202
Mount Vernon Reformed Church Hospital, 139–145
Mumma Farm, 23, 60, 105
Museums
    National Museum of Civil War Medicine, 117, 157, 187–190
    Philip Pry House Field Hospital Museum, 117–127
    Wistar Museum, 135
Myer, Dr Albert J., 155

N

National Museum of Civil War Medicine, 117, 157, 187–190
Nesbit, Colonel, 34
New Protestant Episcopal Church, 180–183
19th Massachusetts Infantry, 119
19th Pennsylvania Infantry, 82
9th New York Infantry, 106, 107, 112, 136, 178
Nongovernmental agencies, 48
North Woods
    battle, 21
    field hospital, 22–23
    map, 26
    wounds and injuries, 24–25
Nurses role, 48

O

Oden, Lieutenant John, 34
Old German Reformed Church, 175–176
Old Protestant Episcopal Church, 180–183
O'Leary, Dr Charles, 202
132nd Pennsylvania Infantry, 72–73
124th Pennsylvania Infantry, 25
Ostheimer, Private I., 166–167
Otho J. Smith Farm Hospital, 129–131
Otto Farm Hospital, 110–111

P

Palmer, Dr G.S., 120
Parker, Captain William W., 31, 37–38
Parkinson, Lieutenant, 38
Parran, Dr William, 66
Patterson, Corporal Lewis, 104
Peabody, Dr James Henry, 160–161
Pension files, 197–199
Perry, Captain J.B., 33
Peters, Dr D.C., 173–174
Philip Pry House Field Hospital
    Coffin's description of scene, 118–119
    contributions of Dr Benjamin Douglas Howard, 122–124
    mortise and tenon joint within barn, 126
    open air barn housing enlisted wounded, 125–126
    Schell description of scene, 119
    sketch of, 125
    wounds and injuries, 119–122
Philip Pry House Field Hospital Museum
    photograph, 117
    telephone number, 118
    Wheeling-Rosecrans ambulance wagon, 126–127
Pineo, Dr Peter, 202
Piper's Farm, 66
Poffenberger, Samuel, 53–55
Poffenberger Farm
    hospital, 53
    House and Barn, 54–55
    wounds and injuries, 56–60

Porter, General Fitz-John, 5
Presbyterian Church, 177–180
Pry, Philip, 117–127
Pry, Samuel, 127–128
Pry House Field Hospital
  Coffin's description of scene, 118–119
  contributions of Dr Benjamin Douglas Howard, 122–124
  mortise and tenon joint within barn, 126
  open air barn housing enlisted wounded, 125–126
  Schell description of scene, 119
  sketch of, 125
Pry House Field Hospital Museum
  photograph, 117
  telephone number, 118
  Wheeling-Rosecrans ambulance wagon, 126–127
Pry Mill Hospital, 127–128

R
Ransom, Chaplain, 33
Reber, Dr Charles T., 100
Reed, Dr T.B., 186
Reel Farm field hospital, 42
Reformed Academy, 154
Reformed Church, 175–176
Reily, Dr John Theodore, 159
Rib injuries, 79–81
Richardson, Major General Israel B., 32, 34
Rickett, Major General James B., 24
Rivers, Dr Henry Wheaton, 202
Rogers, Private Barney, 73
Rohrbach Bridge. *See* Burnside Bridge
Roman Catholic Novitiate, 161
Rosecrans, Major General William S., 126
Rosecrans ambulance wagon, 126–127
Roulette Barn Hospital, 76–78, 99–102, 131
Roulette Farm, 23, 75–82
Roulette Spring House, 81–82

S
Sacket, Colonel Delos B., 5
Samuel Pry Mill Hospital, 127–128
Sanitary Commission, US, 5–6, 48
Schell, Frank H., 22–23, 119
2nd Georgia Infantry, 105
2nd Maryland Infantry, 99–102
2nd United States Sharpshooters, 42–43
Sedgwick, Major General John, 37, 38, 62, 137
7th Michigan Infantry, 162
71st Pennsylvania Infantry, 38–41
Sexton, Dr Samuel, 131
Sharpsburg, Maryland, 60–66
Sherrick Farm Hospital, 110–111
Sherwood, Dr T.H., 186
Signal system, 155
16th Connecticut Infantry, 109
61st New York Infantry, 73
63d New York Volunteers, 79
64th New York Volunteers, 78
Sloan, Reverend I.O., 137, 153
Smith, Dr Otho J., 129–131
Smith, Dr Otto J., 110
Smith, Private J.O., 77–78
Smith Farm Hospital, 129–131
Smith, Nathan R., 193
Smoketown Hospital, 50–51
Sneden, Robert Knox, vi
Sons of Erin, 63
South Mountain, Battle of, 85, 153, 154
Spencer, Private Orrin C., 98
Squires, Dr Truman H., 106, 107
Stickley, Lieutenant Ezra, 29–30
Stone School House, 143
Storrs, Dr Melancthon, 109
Straw-stack hospitals, 67–74
Stuart, General Jeb, 38
Sumner, Major General Edwin V., 37, 62
Sunken Road, Sharpsburg, Maryland, 60–66, 131
Surgeon Generals of the US Army, 2–3
Sweitzer, Lieutenant Colonel Nelson B., 5

Index **209**

T

Taylor, Dr John Howard, 34, 202
Telephone numbers
    National Museum of Civil War Medicine, 118, 157
    Philip Pry House Field Hospital Museum, 118
3rd Wisconsin Infantry, 32, 60
13th New Jersey Infantry, 77
35th Massachusetts Infantry, 102–104
33d New York Infantry, 70
Thompson, Private Franklin Flint, 143–144
Toombs, Brigadier General Robert A., 105
Trinity Chapel, 175
Tripler, Major (Dr) Charles S., 6, 84, 88
Tuberculosis, 96
21st Connecticut Infantry, 96
Typhoid fever, 96

U

United States Christian Commission, 137, 153
United States Sanitary Commission, 5–6, 48, 137–138, 142–143
US Hotel Hospital, 160

V

Valley Mills. *See* Philip Pry House Field Hospital; Samuel Pry Mill Hospital
Vanderveer, Dr A., 197, 198
Visitation Academy, 161–167
Visual signal system, 155
Vollum, Lieutenant Colonel Edward P., 202

W

Ward, Private Morris, 79–82
Watson, Dr Alexander T., 202
Webb, Colonel Alexander S., 5
Weir, Dr R.F., 165
Welles, Cornelius M., 56
Wheeler, Assistant Surgeon William B., xx–xxi
Wheeling-Rosecrans ambulance wagon, 126–127
White, Dr William J.H., 202
Whittier, John Greenleaf, 168
Wier, Dr R.F., 79
"Wigwag" system, 155
Willard, Dr J.N., 120
Willard, Private Van R., 32, 134
Wilson, Lieutenant, 134
Winchester, Hiram, 170, 177
Winchester Hall, 171
Winchester Seminary, 170–172
Windage, 101
Wistar, Brigadier General Isaac J., 38–41, 134–135
Wistar, Dr Casper, 135
Wistar Museum, 135
Woden, Lieutenant John, 34
Wounds and injuries. *See also* Field hospitals
    alcohol abuse, 73
    amputations, 139–142, 163–165, 193–195
    Battle of Antietam casualties, 200
    battlefield experiences of Clara Barton, 56–60
    Bloody Lane, 62–63, 72–73
    bone fractures, 193–194, 196
    brain injuries, 197–199
    Burnside Bridge, 98–109
    casualties by phase of battle, 200
    casualties by type, 200
    cause of, xix
    diarrhea, 37
    Dunker Church, 38–43
    evacuation and treatment of wounded, 74, 149–153
    General Hospital Camp A, 188
    General Hospital Number 1, 187–188, 193–194
    General Hospital Number 3, 183–186
    General Hospital Number 4, 172–174
    General Hospital Number 5, 162–167
    Hawkins' Zouaves, 112–114
    head wounds, 42–43
    Keedysville Hospitals, 133–139, 144–145

Meagher's Irish Brigade, 71–72
Miller Farm Cornfield, 28–32
North Woods battle, 24–25
number of casualties in worst Civil War battles, 201
Philip Pry House Field Hospital, 119–122
rib injury, 79–81
Roulette Farm, 77–82
Smoketown Hospital, 50–51
Sunken Road, Sharpsburg, Maryland, 65–66
survival of, xix
windage, 101
Wunderlich, George C., 62, 63, 84–85, 103

Z

Zacharias, Reverend Dr Daniel, 175–176
Zion Lutheran Church Hospital, 153
Zouaves, 106, 107, 112, 136, 178